D1327302

Natural Computing Series

Series Editors: G. Rozenberg
Th. Bäck A.E. Eiben J.N. Kok H.P. Spaink

Leiden Center for Natural Computing

Anthony Brabazon · Michael O'Neill

Biologically Inspired Algorithms for Financial Modelling

With 92 Figures and 39 Tables

 Springer

Authors

Dr. Anthony Brabazon
University College Dublin
Belfield, Dublin 4
Ireland
anthony.brabazon@ucd.ie

Dr. Michael O'Neill
University College Dublin
Belfield, Dublin 4
Ireland

Series Editors

G. Rozenberg (Managing Editor)
rozenber@liacs.nl

Th. Bäck, J.N. Kok, H.P. Spaink
Leiden Institute of Advanced
Computer Science
Leiden University
Niels Bohrweg 1
2333 CA Leiden, The Netherlands

A.E. Eiben
Vrije Universiteit Amsterdam
The Netherlands

Library of Congress Control Number: 2005936099

ACM Computing Classification (1998): F.1, F.2, I.2.1, I.2.8, I.6, J.1, J.4

ISBN-10 3-540-26252-0 Springer Berlin Heidelberg New York
ISBN-13 978-3-540-26252-7 Springer Berlin Heidelberg New York

Springer is a part of Springer Science+Business Media

springer.com

© Springer-Verlag Berlin Heidelberg 2006
Printed in Germany

Cover Design: KünkelLopka, Werbeagentur, Heidelberg
Typesetting: by the Authors
Production: LE-TeX Jelonek, Schmidt & Vöckler GbR, Leipzig
Printed on acid-free paper   45/3142/YL – 5 4 3 2 1 0

To Maria
Tony

To Gráinne
Michael

Preface

The field of biologically inspired computing has advanced rapidly over the past decade. One offshoot of this progress has been the development of a large family of biologically inspired algorithms. Broadly speaking, biologically inspired algorithms draw metaphorical inspiration from diverse sources, including the operation of biological neurons, processes of evolution, models of social interaction amongst organisms, and natural immune systems, in order to develop tools for solving real-world problems. This book provides an introduction to a broad range of biologically inspired algorithms and illustrates how they can be applied for financial modelling using a series of case studies. These cases include the modelling of financial markets, the development of financial trading systems, the creation of solvency prediction systems, and the creation of credit rating models. In this book particular emphasis is placed on evolutionary methodologies, particularly the novel, powerful, modelling methodology Grammatical Evolution. No prior knowledge of either biologically inspired algorithms or financial modelling is assumed.

We hope that this book will help spark new ideas in the minds of readers to encourage them to undertake their own work in financial modelling using biologically inspired methodologies.

Anthony Brabazon
October 2005 *Michael O'Neill*

Acknowledgment

We would like to thank several people who contributed to the writing of this book. Thomas Randles and David Brennan helped collect the corporate failure and bond-rating datasets used in some of the case studies. We also thank John McCallig, Sean McGarraghy and David Edelman for useful discussions on several aspects of the material in the text which helped clarify our thoughts (all errors are of course our responsibility!). In particular we would like to thank a number of people who co-authored some of the case study chapters with us: Ian Dempsey who co-authored the adaptive trading case study (Chapter 14); Peter Keenan, Katrina Meagher and Edward Carty who were co-authors of the intra-day case study (Chapter 15); Yue Xi and Qiang Han who were co-authors of the corporate failure ant-model case study (Chapter 18); and Peter Keenan, Alice Delahunty and Denis O'Callaghan who contributed to the AIS bond-rating chapter (Chapter 20). We also extend our thanks to Ronan Nugent of Springer-Verlag for his encouragement of this project, and for his advice on early drafts of the manuscript.

Anthony Brabazon
Michael O'Neill

Contents

Part II Model Development

Part III Case Studies

1

Introduction

Over the last decade, a considerable literature on biologically inspired algorithms (BIA) has emerged. These powerful algorithms can be used for prediction and classification, and have clear application for use in financial modelling and in the development of trading systems. Financial markets represent a complex, ever-changing, environment in which a population of investors are competing for profit. Biological entities have long inhabited such environments, and have competed for resources to ensure their survival. It is natural to turn to algorithms which are inspired by biological processes to tackle the task of survival in a financial jungle.

The primary objectives of this book are twofold: to provide readers with an up-to-date introduction to a broad range of BIAs, and to illustrate by means of a series of case studies how the algorithms can be applied for the purposes of modelling financial markets, for the prediction of corporate failure, and for the prediction of credit ratings. Although we cannot provide any guarantees that these technologies provide an easy route to financial riches, we hope this book will spark new ideas in the minds of readers to encourage them to undertake their own work in the fascinating nexus of computer science and finance.

This book is aimed at two audiences: those in the finance community who wish to learn about advances in biologically inspired computing and how these advances can be applied to financial modelling; and those in the computer science community who wish to gain insight into the domain of financial modelling and trading system design. Strong emphasis is placed in this book on evolutionary methodologies, particularly *Grammatical Evolution* [174]. This book is also suitable for use on advanced undergraduate or postgraduate courses, on quantitative finance or computational intelligence. No prior knowledge of either BIAs or financial prediction is assumed.

1.1 Biologically Inspired Algorithms

Biological systems are a notable source of inspiration for the design of optimisation and classification algorithms, and all of the methodologies in this book have their metaphorical roots in models of biological and social processes. These processes are as diverse as the operation of the central nervous system, biological evolution, the mapping of genes to proteins, the human immune system, and models of social interaction amongst organisms. BIAs do not seek to perfectly imitate the complex workings of these systems, rather they draw metaphorical inspiration from them to create mathematical algorithms which can be used in an attempt to solve hard, real-world problems, such as modelling financial markets. Figure 1.1 provides a broad taxonomy of some of the primary methodologies discussed in this book. A vast number of hybrid models which combine elements from more than one of these methodologies can also be constructed.

It is not possible to undertake a complete discussion of all of these in a single text, and we concentrate on neural network and evolutionary algorithms, while providing an introduction to BIA technologies drawn from social and immune metaphors. A brief overview of some of these technologies is provided in the following paragraphs, with a more detailed discussion of them being provided in later chapters.

1.1.1 Artificial Neural Networks

Artificial neural networks (NNs) is a modelling methodology whose inspiration arises loosely from a simplified model of the workings of the human brain. Both learn from their environment and encode this learning by altering the connections between individual processing elements, neurons in the case of the human brain, nodes in the case of NNs. NNs can be used to construct models for the purposes of prediction, classification and clustering. NNs are a non-parametric modelling tool, as the model is developed directly from the data.

1.1.2 Evolutionary Computation

Evolutionary algorithms draw inspiration from the processes of biological evolution to *breed* solutions to problems. These problems may be as diverse as determining the coefficients for a non-linear regression model, or determining the components of a financial trading system. The algorithm commences by creating an initial population of potential solutions, and these are iteratively improved over many 'generations'. In successive iterations of the algorithm, fitness-based selection takes place within the population of solutions. Better solutions are preferentially selected for survival into the next generation of solutions, with diversity being introduced in the selected solutions in an attempt to uncover even better solutions over multiple generations. BIAs that

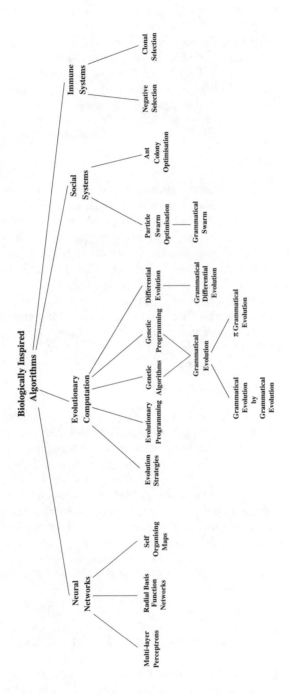

Fig. 1.1. A taxonomy of the biologically inspired algorithms that are discussed in this book

employ an evolutionary approach include genetic algorithms (GAs), genetic programming (GP), evolutionary strategies (ES) and evolutionary programming (EP).

A significant recent addition to BIA methodologies is grammatical evolution (GE), an evolutionary automatic programming methodology, which, for example, can be used to evolve rule sets or financial trading systems. GE incorporates a *grammar* which governs the creation of these rule sets. The idea of a grammar is inspired by the biological process of the mapping of genes to proteins.

1.1.3 Social Systems

The social models considered in this book are drawn from a *swarm* metaphor. Two popular variants of swarm models exist, those inspired by the flocking behaviour of birds and fish, and those inspired by the behaviour of social insects such as ants. The essence of these systems is that they exhibit flexibility, robustness, self-organisation, and communication between individual members of the population. The swarm metaphor has been used to design algorithms which can solve difficult problems by creating a population of problem-solvers, and allowing these to communicate their relative success in solving the problem to each other. Higher-performing individuals attract the attention of others, who test variants on their problem-solving strategy in an attempt to improve it.

1.1.4 Artificial Immune Systems

The human immune system is a highly complex system, comprised of an intricate network of specialised tissues, organs, cells and chemical molecules. The capabilities of the natural immune system are to recognise, destroy and remember an almost unlimited number of foreign bodies, and also to protect the organism from misbehaving cells in the body. To assist in protecting the organism, the immune system has the capability to distinguish between *self*, and *non-self*. Artificial immune systems (AIS) draw inspiration from the workings of the natural immune system to develop algorithms for optimisation and classification. Practical applications of AIS models to pattern-recognition tasks include the identification of potentially fraudulent credit card transactions, the identification of the 'state' of the stock market, and the identification of financially at-risk companies.

1.2 Computer Trading on Financial Markets

Computerised or automated trading on financial markets is not a new phenomenon. Computers have been used for *program trading* for many years. In program trading, the object is usually to uncover and eliminate anomalies between financial derivatives and the underlying financial assets which make up those derivatives.[1] A typical example of program trading is *index arbitrage* which involves the automated purchase or sale of a basket of stocks which make up a market index, in conjunction with the simultaneous sale or purchase of a derivative product such as stock index futures, in order to profit from the price difference between the basket and the derivative product. In theory the transaction generates risk-free returns, but in practice it relies on estimates of dividend income from companies, an estimate of the rate of return available on invested dividends, and the ability of the computer to make the purchases/sales at the prices which produced the arbitrage opportunity. Program trading accounts for a considerable portion of trading on major stock exchanges. For example, it is estimated that program trading volume accounted for approximately 50.6% of the total trading volume on the New York Stock Exchange (NYSE) in 2004 [163].

A second, less publicised use of computers is to construct trading systems which assume trading risk in the search for superior, risk-adjusted, returns. These systems are the focus of interest of several of the case studies in this book.

1.3 Challenges in the Modelling of Financial Markets

Modelling of financial markets is challenging for several reasons. Many factors plausibly impact on financial markets including interest rates, exchange rates, and the rate of economic growth. We have no hard theory as to how exactly these factors effect prices of financial assets, partly because the effects are complex, non-linear and time-lagged. For example, a change in interest rates may impact on the foreign exchange rate for a currency, in turn effecting the level of imports and exports into and from that country. Another difficulty that arises in financial modelling is that unlike the modelling of physical systems we cannot conduct controlled experiments. Only one sample path through time is available for our examination, as we only have one history of market events. Additionally, some factors which can effect financial markets are inherently unpredictable such as earthquakes, the weather, or political events. Taken together, these difficulties imply that our ability to predict market movements will always be imperfect.

---

[1] A derivative is a financial instrument whose value is based on that of another financial instrument such as a share. For example, investor A may sell an option on a share to investor B. This option gives investor B the right to buy (or sell) that share to investor A, at a specified price for a specified time. As the value of the underlying share changes, the value of the financial derivative (the option) will also change.

1.3.1 Do Prices Follow a Random Walk?

The very attempt at modelling financial markets for profit meets with the scorn of some financial economists. Two pillars of traditional financial economics are that market prices of financial assets follow a *random walk* and that markets are *efficient*. One of the earliest studies suggesting that prices in markets might follow a random walk was undertaken by [123]. The traditional definition of a random walk is a process in which the changes from one time period to the next are independent of each other, and are identically distributed. If prices in financial markets did follow a random walk, this would imply that the size and direction of a past change in price provides no insight into the size and direction of the next change in price of that financial asset. In other words, there would be no auto-correlation in the time-series of prices from a financial market.

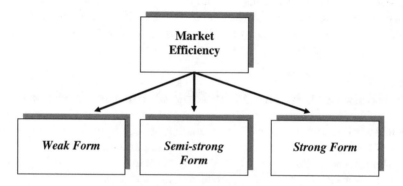

Fig. 1.2. Three forms of market efficiency

Closely related to the concept of a random walk (and often confused with it) is the efficient market hypothesis (EMH). Although a random walk in share prices could arise for a variety of reasons, it is consistent with a proposition that current prices fully reflect the market's aggregate assessment of any existing information which could impact on the price of a financial asset. If a market is informationally efficient, in that all information is impounded accurately and instantly into prices once it becomes available, then there is no scope to make excess returns from trading on such information. The more information-efficient a market is, the more random the sequence of price changes that it will produce, as prices will only alter when new information emerges. As the nature of new information is by definition unpredictable, a share's price is equally likely to rise or fall in the future. Three versions of market efficiency were posited [73] (Fig. 1.2).

Under weak form efficiency it is considered that the price of a share at any point in time reflects *all* the information contained in its price history. This would imply that excess risk-adjusted returns cannot be obtained by attempting to construct a model which uses information on past share prices, or past transaction volumes, to predict future share prices. The semi-strong form of efficiency suggests that a share price at any point in time reflects all publicly available information which could impact on the share's price. The strong form of market efficiency implies that share prices include all information relevant to the price of a share, including both public and private (non-public) information.

1.3.2 Attack of the Anomalies

If the semi-strong form of the EMH was correct, there would be no scope to construct a model of a financial market using publicly available information, which would generate excess risk-adjusted returns. In spite of the initial research which lent broad support to the EMH, there is a growing body of research in more recent times which suggests that subtle patterns do exist in time-series of financial asset prices, and that these prices do not follow a random walk [143, 144]. Among the anomalies that have been noted are the existence of serial correlation in weekly and monthly stock returns, particularly for small capitalisation (small company) stocks. Generally, three patterns of serial correlation in stock returns are recognised: short-term reversals (looking at returns over a few weeks), medium-term inertia, and longer-term reversals.

A considerable body of empirical evidence suggests that short-run volatility in share returns is clustered. A large change in price tends to be followed by another large change in price, but the direction of this change is difficult to predict. In other words, prices tend to be volatile when they have just been volatile, leading to patterns of short-run price reversals. Under medium-term inertia, good (or bad) performance of a stock over 3-12 months is typically indicative of continued good (or bad) performance in the next few months. This could provide scope for the implementation of *momentum* investment strategies, where investors seek to buy (sell) shares which have recently trended upwards (downwards). Over longer time periods (3-5 years), there is evidence of negative serial correlation in share returns, suggesting that stocks that have performed well over the last several years are more likely to under-perform in the future. This lends support to the common idea of *contrarian* or *value* investment strategies where investors buy out-of-favour stocks (those whose share price has underperformed that of their peers in recent years). A posited explanation for this negative serial correlation is the *over-reaction hypothesis* that investors are subject to waves of optimism and pessimism which cause prices to swing temporarily away from their underlying value for individual firms or whole sectors [53, 63]. Other examples of asset-price anomalies are reported in [37] and [54].

Interpretation of the results of studies reporting anomalies has been controversial [74, 75], but they are consistent with a hypothesis that market efficiency is a relative term. Under this premise, as market participants uncover new information processing mechanisms (such as BIAs), market efficiency is enhanced as market participants apply the new information processing methodology. It is plausible that novel, powerful computational techniques which can uncover new price-relevant information could prove profitable. However markets represent a competitive, adaptive environment and are likely to rapidly impound available information into asset prices. Just as $100 bills do not last long on the sidewalk, traders who constitute markets have a vested interest in searching for and exploiting any edge which could lead to profit. Financial modelling in such an environment can be compared to an arms race whereby each player rapidly cancels out any advance of another. The advantage offered by a new modelling technique is therefore likely to be short-lived.

There are close parallels between the challenges of the environment of financial markets, and those from which several of the biologically inspired algorithms discussed in this book are drawn. Evolutionary algorithms, swarm algorithms and immune algorithms are drawn from environments where, just as in financial markets, there is continual adaptation and where there is competition for resources between individuals.

1.4 Linear Models

The goal in modelling a system is typically to gain insight into its behaviour, to determine which factors impact on the output of the system, and to determine how influential each of these factors is. A second goal is to enable prediction of the future output of the system under different conditions. A simple linear model has the general form:

$$Y = \alpha + \beta_1 X_1 + \ldots + \beta_n X_n \tag{1.1}$$

where Y is the dependent variable, $X_1, \ldots, X_n$ are independent (or explanatory) variables (in a simple model there may be only one independent variable, in a multiple regression model there will be several), $\beta_1, \ldots, \beta_n$ are regression coefficients, and α is a constant which allows the model to produce a value for Y even if all the dependent variables have a zero value.

In constructing a linear model, the first step is typically to posit a *cause-and-effect* relationship based on prior theory or intuition between the dependent variable and one or more explanatory variables (or inputs). In other words: 'I think x and y impact on the value of z'. Two questions then spring to mind:

i. Is the assumption that x and y effect z supported by empirical evidence?
ii. How great is the effect of x and y respectively on z?

To answer these questions, we can collect sample data, vectors of explanatory variables and the associated value of the dependent variable, and then either manually or using a computer package determine the values for the regression coefficients which produce a model whose output closely agree with actual known output for each vector of input data. If values can be found for the regression coefficients such that the linear model is successful in explaining a high portion of the variation in the dependent variable (z), where the signs (+ or −) of the regression coefficients concord with theory/intuition, we consider that the model is good, and that our hypothesis that x and y impact on z is plausibly supported by the collected data.

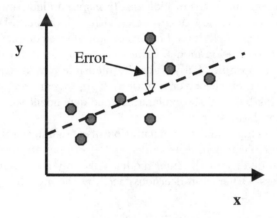

Fig. 1.3. The least-squares regression line is constructed by minimising the sum of the squared errors for each data-point. The error for each point is the difference between its actual value of y, and the predicted value of y according to the regression line

The Error Measure

To determine the values of the regression coefficients, the modeller must define an error measure so that the error between the model's predicted output and the actual output value can be calculated. This error is then used as feedback to alter the regression coefficients in order to reduce the error measure. Typically in basic regression models the goodness-of-fit measure is the sum of the squared errors between the predicted and the actual outputs (Fig. 1.3). If we are willing to make assumptions concerning the distribution of the error terms resulting from the model's predictions, i.e. that the errors

- have a mean of zero,
- are normally distributed,
- are independent, and
- have constant variance,

then a series of statistical statements can be made including the construction of confidence intervals for the model's predictions and for the values of the regression coefficients. One issue of particular interest is whether the regression coefficients are significantly different from zero.

Modelling with Biologically Inspired Algorithms

In applying the various forms of biologically inspired algorithms, we are undertaking the same basic modelling process, although the actual mathematical form of the resulting model may be considerably more complex than that of the simple linear regression model. If we want to predict a future share price or other financial variable, can we identify plausible sets of explanatory (input) variables based on theory or intuition? If so, we can test our hypothesis that there is a link between the explanatory and dependent variables by using historical market data.

Although the choice and implementation of modelling methodology (linear regression, artificial neural networks, etc.) can play an important role in determining the quality of the final model, it is only one component of the modelling process. Other vital decisions faced by the modeller address questions such as:

- What data should be used to construct the model?
- Is the cause-and-effect relationship plausible?
- Does this data need to be preprocessed before it is included in the model?
- What error measure is appropriate?

No modelling methodology will compensate for poor decisions in these steps, and each of these issues will be discussed in later chapters.

1.5 Structure of the Book

The remainder of this book is divided into three parts. In Part I a range of biologically inspired algorithms are introduced and explained. These offer the potential to develop useful financial models. However, despite the undoubted power of these algorithms, their successful implementation requires the careful selection of explanatory variables, and the careful validation of the results arising from the developed models. Therefore the book contains a section on model development (Part II), which covers a range of practical issues which arise in the creation of financial models. Finally, in Part III, a series of case studies are provided to illustrate several potential financial applications of biologically inspired methodologies. A number of these cases concentrate on the construction of trading systems in equity and foreign exchange markets. The utility of the methodologies is further demonstrated through their application to a range of other tasks, including the prediction of corporate failure, and the prediction of corporate bond ratings.

Part I

Methodologies

2

Neural Network Methodologies

The human brain consists of about 100 billion nerve cells or *neurons*. Each of these is interconnected to a few thousand other neurons, and is constantly receiving electrical signals from them along fibres, called dendrites, that emanate from the cell body. If the total signal coming into an individual neuron at a point in time exceeds a threshold value, the neuron fires and produces an outgoing signal along its axon, which in turn is transmitted to other neurons. Connections between neurons occur at synapses, and signals cross the synaptic gap by means of a complex electro-chemical process. The brain can therefore be stylised as a vast interconnected, parallel-processing unit. This unit receives inputs from its environment, it can encode and recall memories, and it can integrate inputs to produce a thought or an action (an output). The brain has the capability to recognise patterns, and to predict the likely outcome of an event based on past learning.

Artificial neural networks (NNs) comprise a family of mathematical modelling methodologies whose metaphorical inspiration is loosely drawn from the workings of the human brain and central nervous system. NNs can be used for a wide variety of tasks including the construction of models for the purposes of prediction, classification and clustering.

With the wide availability of both historical financial data and commercial NN software, there has been an increasing volume of literature applying NNs to prediction, of both equity markets [13, 76, 86, 186, 226] and foreign exchange markets [82, 112, 115, 223]. NNs have also been extensively applied in other financial applications, notably credit rating, prediction of financial distress, and fraud detection. Wong, Lai and Lam [221] provide a comprehensive bibliography of the literature on business applications of NNs.

2.1 A Taxonomy of NNs

A wide variety of NN architectures and training algorithms exist. These can be differentiated from each other along three main axes:

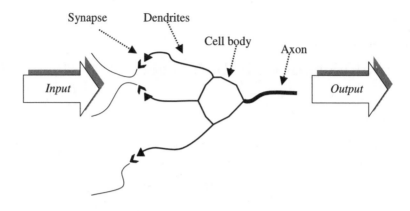

Fig. 2.1. A simplified diagram of a nerve cell

   i. connection topology,
  ii. training method, and
 iii. learning algorithm.

The connection topology defines how the processing units or nodes are connected to each other. The training method is concerned with how the NN learns. In supervised learning, the NN is provided with training data (input data for which the output is already known), and over multiple iterations of the learning algorithm discovers how to link the inputs to the associated known outputs. Unsupervised learning occurs when the NN is not provided with outputs, but rather is left to uncover patterns in the input data. An example of an unsupervised learning scenario would be the uncovering of previously unknown patterns in databases of customer information. The learning algorithm defines how error is measured during the training process, and how the NN model is updated during training in order to reduce this error.

Many forms of NNs can be developed by making different choices for the above three items. Three common NN structures (Fig. 2.2) are described in the following sections: the multi-layer perceptron, radial basis function networks, and self-organising maps.

2.2 The Multi Layer Perceptron

The canonical NN model, the multi-layer perceptron (MLP), consists of a multi-layer, feedforward architecture, and is trained using the backpropagation training algorithm. The architecture is described as feedforward as the pattern of activation of the network flows in one direction only, from the in-

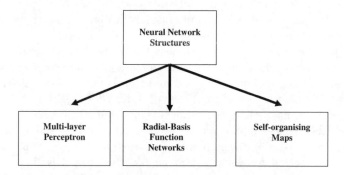

Fig. 2.2. Common neural network structures

put to the output layer. MLPs usually consist of three layers of interconnected computing units called nodes (Fig. 2.3).

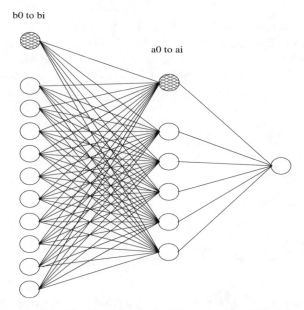

Fig. 2.3. An example of a three-layer feedforward MLP. The left-most layer of nodes is the input layer, the rightmost node is the output node, and the middle layer of nodes is the hidden layer

The first layer, the *input layer*, serves as a holding layer for the data being input to the MLP. A vector of input data is presented to the input layer (*node_i* in the input layer receives *element_i* in the data vector). This layer is

connected to one or more *hidden layers* (so called as they are not directly
connected to the outside world), and nodes in this layer are connected in turn
to an *output layer* which represents the processed output from the model.
Each of the connections (*arcs*) between the nodes has an associated real-
valued *weight*, and this weight is similar in concept to a regression coefficient.
The signal or value passing along a connection is modified by multiplying it
by this weight before it reaches the next node. The weight therefore serves
to amplify or dampen the strength of a signal (value) passing along an arc.
Generally, the processing carried out at each node in the hidden and output
layers consists of passing the sum of the weighted inputs to that node through
a non-linear function, known as a *transfer function* (Fig. 2.4). Typical choices
for the transfer function are logistic or hyperbolic tan (tanh) functions, which
transform an input in the range $-\infty$ to $+\infty$ to the range (0,1) and (-1,1)
respectively (Fig. 2.5). The logistic function has the form

$$y_j = \frac{1}{1 + \exp^{-\Sigma}} \qquad (2.1)$$

where $\sum$ is the weighted sum of the inputs into $node_j$, and y_j is the output
from $node_j$.

It is easily demonstrated that a simple 2-layer MLP with a single output
node, no hidden layer and a linear transfer function is equivalent to a linear
regression model, where the arc weights correspond to regression coefficients.
For example, the regression equation $Y = a + bx_1 + cx_2$ can be represented
as in Fig. 2.6. Similarly, a logistic regression model can be recast as a 2-layer
MLP with a sigmoid transfer function. A multi-layer MLP can therefore be
described as a non-parametric, non-linear regression model [98].

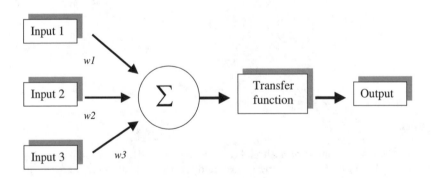

Fig. 2.4. A single processing node in an MLP. The arc weights are denoted $w1$, $w2$
and $w3$. The weighted sum of the inputs to the node is passed through a transfer
function, to produce the node's output

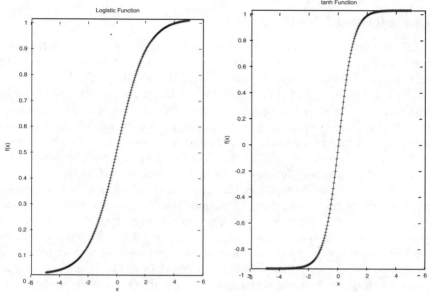

Fig. 2.5. Logistic and tanh functions

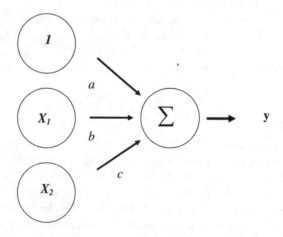

Fig. 2.6. Linear equation as a node-arc structure $(y = a + bx_1 + cx_2)$

Characterising a MLP

MLPs provide an example of parallel distributed computing, as each hidden layer node acts as a local processor of information, yet also acts concurrently and in parallel with the other nodes in its layer. Although the processing which takes place at individual nodes is relatively simple, the linkage of individual nodes gives rise to emergent capabilities for the network, permitting complex non-linear, input-output mappings. In essence, a MLP implements a function f that maps a vector of input values $\mathbf{x}$ to one or more output values y: $y = f(\mathbf{x})$. In constructing a MLP the object is to approximate the underlying (but initially unknown) function f as well as possible. MLPs have universal approximator capabilities [46] in that under general conditions they are capable of mapping any continuous function [110, 191]. In contrast to ordinary least-squares regression models which produce a line, plane or hyperplane depending on the number of independent variables, MLPs which use non-linear transfer functions can produce complex (but smooth) response surfaces with peaks and troughs in n dimensions. Changing the weights during the learning process tunes this response surface to more closely fit the training data.

NNs are inductive, *data-driven* modelling tools, which can map non-linear data structures without requiring an explicit a priori specification of the relationship between model inputs and outputs. This is a particular advantage when applied to financial data, as in many cases we have a weak understanding of the causal relationships between variables.

The behaviour of MLPs, how they map inputs to output(s), is influenced primarily by the form of processing that takes place at hidden and output layer nodes, how the nodes are interconnected and how the weights are associated with these interconnections. The general form of the three-layer MLP is:

$$z_t = L \left(a_0 + \sum_{j=1}^{x} w_j L \left(\sum_{i=0}^{y} b_i w_{ij} \right) \right) \tag{2.2}$$

where b_i represents *input$_i$* (b_0 is a bias node), w_{ij} represents the weight between input node$_i$ and hidden node$_j$, a_0 is a *bias* node attached to the output layer, w_j represents the weight between hidden node$_j$ and the output node, z_t represents the output produced by the network for input data vector (t), and L represents a non-linear transfer function. The inclusion of a bias node serves a similar purpose to the inclusion of a constant term in a regression equation. The input value of the bias node is usually held constant at one, and is automatically rescaled as necessary as the weights on its outgoing connections change. The optimal size of the hidden layer is not known a priori and is determined heuristically by the modeller.

2.2.1 Training an MLP

MLPs are trained using a *supervised* learning paradigm. In supervised learning a set of input data vectors for which the output is already known are presented to the MLP. The MLP predicts an output for each input vector, and an error between the predicted and the actual value of the output is calculated. The weights on each connection in the network are then adjusted in order to reduce the error. By altering the weights, the network can place different emphasis on each input, and differing emphasis on the output of each hidden layer node in determining the final output of the network. The *knowledge* of the network is therefore embedded in its connection weights.

The Backpropagation Algorithm

The most common way of altering the weights in response to an error in the network's prediction is through use of the *backpropagation algorithm*. At the start of the learning process the weights on all arcs are initialised to small random values. The network is presented with an input data vector and then proceeds to predict a value for the output. Total squared error is defined as:

$$E = \sum_{p=1}^{P} \sum_{i=1}^{S} (t_i^p - o_i^p)^2 \tag{2.3}$$

where P is the number of input-output vectors, S is the number of output neurons, and t_i^p and o_i^p are the actual and the predicted values of the output.

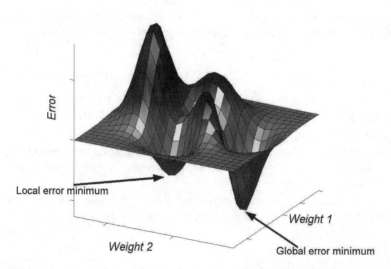

Fig. 2.7. An error surface. As the values of the two weights changes, the resulting network error also changes

The backpropagation algorithm seeks to reduce the total error by calculating the gradient of the error surface at its current point (corresponding to
the current weight vector for the network), and adjusting the weights in the
network in order to descend the error surface (Fig. 2.7). This is achieved by
making a backward pass through the network, from the output to the input
layers, in which weight changes are propagated back through the arcs of the
network, so as to make the prediction of the network better agree with the
actual output value(s). The bigger the error, the more the arc weights are
adjusted. This step is performed repetitively over the entire training dataset,
until learning stops and the network reaches a stable minimum error. Once
the network has been trained, it can be used to predict an output for an input
data vector which it has not previously seen. A simple algorithm for training
a MLP is:

i. Initialise the values of the weights on each connection to small random
   values in the range 0-1.
ii. Present an input vector $\mathbf{x}$: $x_0, x_1, ..., x_{n-1}$ and the associated target output
    O. Assume the network has n input nodes, and that the weights between
    nodes i and j are given by w_{ij}.
iii. Calculate the output from each node in the hidden layer, and then in the
     output layer. The output from a node j in the hidden or output layer is
     given by

$$z_j = \emptyset \sum_{i=0}^{n-1} w_{ij} x_i \qquad (2.4)$$

where $\emptyset$ is the transfer function for that node.
iv. Adjust the weights on the connections between the nodes, commencing
    with the output layer and working back to the input layer as follows:

$$w_{ij}(t+1) = w_{ij}(t) - \alpha \left(\frac{\partial E}{\partial w_{ij}(t)} \right) \qquad (2.5)$$

where $w_{ij}(t)$ is the weight between nodes i and j at iteration t, $\alpha(> 0)$
is the *learning rate*, and $\frac{\partial E}{\partial w_{ij}(t)}$ is the contribution of that weight to the
total network error.[1] Weights may be updated in *batch mode* based on
the total error when all the input-output training data is passed through
the model, or may be updated after each individual training vector is
presented to the network.
v. Repeat steps (ii) - (iv) until the error between the predicted value and the
   actual value reaches a steady state or an acceptable minimum.

---

[1]The total error of the network is a function of the values of all of its weights. In
altering the individual weights during the backpropagation step, in order to minimise
this error, we consider the partial derivative of the total error with respect to each
individual weight $w_{ij}(t)$.

If the above algorithm is applied, there is a good chance that *overtraining* will occur, and that the MLP will start to model noise in the training dataset giving rise to poor performance when applied to new data. The network will not *generalise* well. One way of reducing this problem is to use the method of *early-stopping*. In this method, the dataset is divided into three components: *training* data, *validation* data, and *out-of-sample* (test) data. The MLP is constructed using the training dataset, but periodically during this process, the performance of the network is tested against the validation dataset. The network's performance on the validation dataset is used to determine when the learning process is stopped, and the best network is defined as that which produces the minimum error on the validation dataset. Once the best network is found the weights are fixed and the network is ready for application to out-of-sample data.

2.2.2 Practical Issues in Training MLPs

A number of practical problems arise in using MLPs for modelling.

- What measure of error should be used?
- What parameters should be chosen for the backpropagation algorithm?
- How many hidden layers (or nodes in each hidden layer) should there be?
- Is the data of sufficient quality to build a good model?

Measure of Error

Many different error criteria can be applied in determining the quality of fit of a NN model. Most applications use traditional criteria drawn from statistics such as the sum of the squared errors, or mean squared error (MSE):

$$MSE = \frac{\sum_{i=1}^{n} ||Y_{predict,i} - Y_{actual,i}||^2}{n} \qquad (2.6)$$

where $Y_{predict,i}$ is the output value predicted by the NN model for input vector i, $Y_{actual,i}$ is the actual output value for the input vector, and there are n input-output data vectors. Although this is a common error metric it can lead to poor generalisation, as one way of reducing MSE is to build a large NN which learns the noise in the training dataset. This can be discouraged by using a *regularised* performance function (error measure), where the performance function is extended to include a penalty term which gets larger as network size grows. As an example, the MSE error term could be adjusted to give:

$$\text{error measure} = \alpha \text{ MSE } + (1 - \alpha) \text{ MSW} \qquad (2.7)$$

where MSW is a penalty term, calculated as the mean sum of the squared weights in the network. The values of α and $(1-\alpha)$ represent the relative importance that is placed on the MSE and the penalty term respectively. The penalty term will tend to discourage the use of large weights in the network, and will tend to smooth the response of the network.

Although statistical measures of error are commonly used when developing NNs, it should be remembered that they are not an ideal error measure if the aim is to develop a market-trading system. The objective of a trading system is not to minimise a statistical measure of error, but to maximise risk-adjusted profit.

Parameters for the Backpropagation Algorithm

The essence of training a MLP is the determination of good values for the individual weights in the network. If there are n weights in the network, the task of uncovering good weights amounts to a non-linear optimisation problem where an error surface exists in $(n + 1)$-dimensional space. Unfortunately, no general techniques exist to optimally solve this problem. The backpropagation training algorithm is a gradient-descent, local search algorithm, which is prone to becoming trapped in local optima on the error surface.[2] A number of steps can be taken to lessen the chance of this happening.

Typically, during the training process, the network weights are altered, based on the current model error and a modeller-tunable parameter (the *learning rate*) which governs the size of weight change in response to a given size of error. Usually the value of the learning rate will decay, from a higher to a lower value over the training run, with fairly rapid learning in the initial training stages, and smaller weight adjustments later in the training run. The object in varying the learning rate during the training process is to enable the NN to quickly identify a promising region on the error surface, and later to allow the backpropagation algorithm to approach the minimum error point in that region of the error surface.

However, there is no easy way to determine a priori what learning rates will produce best results. The learning process will typically have an element of *momentum* built in, whereby the direction and size of weight change at each step is influenced by the weight changes in previous iterations of the training algorithm. Therefore the weight change on iteration $t + 1$ is given by:

$$\Delta w_{ij}(t + 1) = \lambda \Delta w_{ij}(t) - (1 - \lambda)\alpha \left(\frac{\partial E}{\partial w_{ij}(t)} \right) \qquad (2.8)$$

where $\lambda \Delta w_{ij}$ is the momentum term, and α is the learning rate. By varying the value of the momentum coefficient, λ in the range 0 to 1, the importance of the momentum coefficient is altered. Under the concept of momentum, if

---

[2]Many gradient-descent-based alternatives to the backpropagation algorithm can be used to train MLPs. A common alternative is the Levenberg-Marquardt (LM) algorithm [102], which approximates the error of the network with a second-order expression, in contrast to the first-order approximation used by the backpropagation algorithm.

the MLP comes across several weight updates of the same sign, indicating a uniform slope on the error surface, the weight update process will gather momentum in that direction. If later weight updates are of different signs, the effect of the momentum term will be to reduce the size of the weight updates below those which would occur in the absence of the momentum component of the weight update formula. The practical affect of momentum is to implement *adaptive learning*, by speeding up the learning process over uniform regions of the error surface.

The backpropagation learning algorithm can be compared to jumping around an error surface on a pogo stick. If the jumps are too small (corresponding to a low learning rate) the pogo stick jumper could easily get stuck in a local minimum, if the jumps are too large, the pogo stick jumper could overshoot the global minimum error. This analogy also underlines the importance of the initial weight vector. The initial weight vector determines the starting point on the weight-error surface for the backpropagation algorithm. If a poor starting point is chosen, particularly if the learning rate is low, the algorithm could quickly descend into an inescapable local minimum (Fig. 2.8). To reduce the chance that a bad *initialisation* of the weight vectors will lead to poor performance of a MLP, performance should be assessed across several training runs using different initialisations of the connection weights (Fig. 2.9).

Data Quality

The quality of the dataset also plays a key role in determining the quality of the MLP. Obviously if important data is not included, perhaps because it is not available, the results from the MLP are likely to be poor. Another data-related issue is how representative the training data is of the whole dataset. If the training data is not fully representative of the behaviour of the system being modelled, out-of-sample results are again likely to be poor (Figs. 2.10 and 2.11). The dataset should be recut several times to produce different training and out-of-sample datasets, and the stability of the results of the developed MLPs across all of the recuts should be considered.

Selecting Network Structure

Although a three-layer MLP is theoretically capable of approximating any continuous function to any desired degree of accuracy, there is no theory to decide how large the hidden layer needs to be to achieve this [46]. Typically the size of this layer is determined heuristically by the modeller. However as the hidden layer gets larger, the number of degrees of freedom consumed by the model rises, and the amount of data needed to train the MLP increases. Broadly speaking, a degree of freedom is consumed by each weight in the MLP, hence a fully connected 20-10-1 MLP (input layer nodes-hidden layer nodes-output node) contains (20*10) + (10*1) = 210 weights. As a rule of thumb,

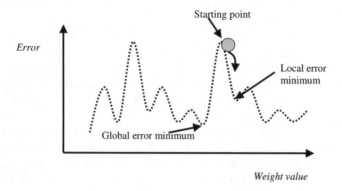

Fig. 2.8. Given this starting point on the weight surface, a gradient-descent algorithm will only find a local error minimum

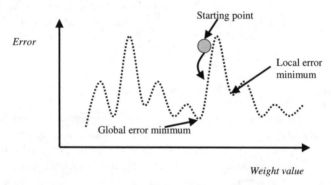

Fig. 2.9. Altering the initial weights moves the starting point on the weight surface, making the global error minimum point accessible

there should be at least 5-10 data vectors for each weight estimated to reduce the chance of overfitting the training data with consequent poor generalisation out-of-sample. Therefore the above network will require a fairly large dataset for training purposes. The selection of the size of the hidden layer entails a trade-off between increasing the power of the MLP (more nodes) and avoiding overfit (fewer nodes).

Sometimes in an attempt to resolve this dilemma, a second hidden layer is added to the basic three-layer MLP, with the size of the original hidden layer being reduced. If the quantity of data for training is constrained, the modeller may try two hidden layers of moderate size, in an attempt to reduce the quantity of training data required while still providing the MLP with reasonable mapping power. For example, if an alternative four-layer MLP was

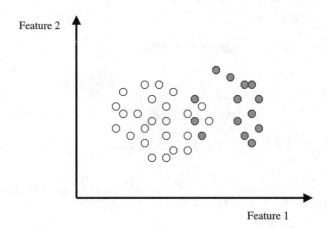

Fig. 2.10. The clear points are training data, and the shaded points are test data. In this cut of the dataset between training and test data, the training data is not representative of the whole dataset

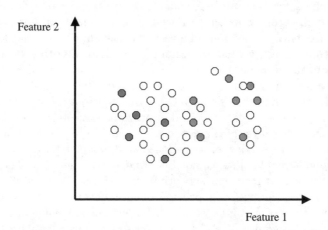

Fig. 2.11. The clear points are training data, and the shaded points are test data. In this recut, the training data is more representative of the whole dataset

proposed for the above dataset with a 20-5-5-1 structure, the total number of weights would drop to (20*5) + (5*5) + (5*1) = 130.

In designing MLPs, there is no restriction that they must have a fully connected feedforward connection structure. Each input need not be connected to each hidden layer node, and nodes can be connected to nodes which are more than one layer ahead in the network (a *jump connection network*) (Fig. 2.12). These networks can be trained using the usual backpropagation method.

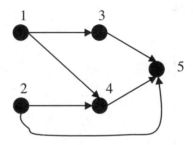

Fig. 2.12. Input 2 is connected to only one hidden layer node, and also has a jump connection directly to the output node

2.2.3 Recurrent Networks

The inspiration for recurrent networks (networks that allow feedback connections between the nodes) is the observation that the human brain is a recurrent network. The activation of a particular neuron can initiate a flow of activations in other neurons which in turn feed back into the neuron which initially fired. The feedback connections in a recurrent network imply that the output from node b at time t can act as as an input into node a at time $t+x$. Nodes b and a may be in the same layer, or node a may be in an earlier layer of the network, and a node may feed back into itself ($a = b$). Recurrent networks can be useful when modelling time-series data, as the recurrent connections allow the network to store information received in previous time steps, and then feed it back into an earlier layer of the network. In contrast, a standard feedforward network has a data window of a fixed size, and associations in the data that extend beyond this window cannot be found by the network. The practical benefit of this is that recurrent network designs can be compact. Consider the case where a modeller wishes to provide a neural network with information on the past N values of M input variables. If a canonical feedforward MLP was used, this would require $M * N$ inputs, possibly a large number, leading to a large number of weights which require training. As recurrent networks can embed a *memory*, their use can reduce the number of input nodes required.

An example of a simple recurrent network is an *Elman network*. This includes three layers, with the addition of a set of *context* nodes which represent feedback connections from hidden layer nodes to themselves (Fig. 2.13). The connections to the hidden layer from these context nodes have a trainable weight. The context nodes act to maintain a memory of the previous period's activation values of the hidden nodes. The output from the network depends therefore on both current and previous inputs. An implication of this is that recurrent networks operate on both an input space and an internal state space. Generalising the context layer concept, it is possible to implement more than one context layer, each with a different lag period. Time is represented implicitly as a result of the design of the network, rather than explicitly through the use of a large number of time-lagged inputs.

Several methods exist to train Elman networks. The original method proposed [71] was to treat each of the feedback inputs from the context layer as an additional input to the network at the next time step, and a standard backpropagation algorithm was used to train all the weights in the network. Other methods of training these networks include the use of backpropagation through time [187]. Training of recurrent networks using gradient-based methods can be time-consuming, and alternative methods using evolutionary, particle swarm or hybrid approaches exist [194].

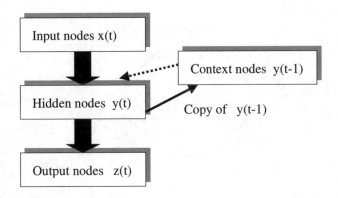

Fig. 2.13. An Elman network. The output of each of the hidden layer nodes at time $t - 1$ is stored in individual context nodes, and each of these are fed back into all the hidden layer nodes as an input, at time t. The context layer nodes are empty during the first training iteration

2.3 Radial Basis Function Networks

A radial basis function network (RBFN) generally consists of a three-layer feedforward network, and is constructed using a supervised training process. Just as for the MLP, the RBFN can be used for prediction and classification purposes, but RBFNs differ from MLPs in that the activation functions of the hidden layer nodes are radial basis functions. The use of RBFNs for classification purposes is based on Cover's theorem on the separability of patterns [44]. This theorem states that complex pattern classification problems are more likely to be linearly separable if the patterns are initially projected nonlinearly into a higher dimensional space. Therefore when using a RBFN for classification purposes, the hidden layer nodes act to project the input vector into a higher dimension (so there should be more hidden layer nodes than inputs) after which it can be classified using a linear transfer function at the output layer.

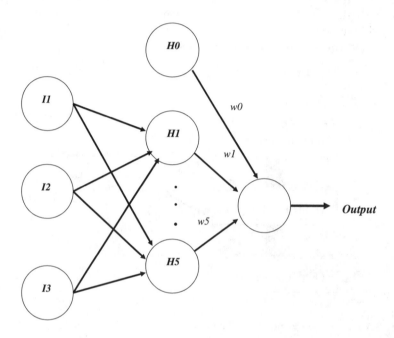

Fig. 2.14. A radial basis function network. The output from each hidden node (H0 is a bias node, with a fixed input value of 1) is obtained by measuring the distance between each input pattern and the location of the hidden node, and applying the radial basis function to that distance. The final output from the network is obtained by taking the weighted sum (using w0, w1 and w5) of the outputs from the hidden layer and from H0

The training of RBFNs typically consists of a combination of unsupervised and supervised learning. Initially, a number hidden layer nodes (or *centres*) must be positioned in the input data space. This can be performed by following a simple rule, or in a more sophisticated application by using unsupervised learning. Methods for choosing the locations of centers include distributing the centres in a regular grid over the input space, selection of a random subset of the training data vectors to serve as centres, or using an algorithm to cluster the input data (SOMs, which are described in the Sect. 2.4 can be used for this) and then selecting a centre location to represent this cluster. Each of these centres forms a hidden node in the RBFN's structure.

Input data vectors are typically standardised before training. When each input vector is presented to the network a value is calculated at each centre using a radial basis function. This value represents the quality of the match between the input vector and the location of that centre in the input space. The greater the distance between an input vector and a particular hidden node, the lower the activation value of the node. Each hidden node, therefore, can be considered as a local detector in the input data space. The most commonly used radial basis function is a Gaussian function. This produces an output value of one if the input and weight vectors are identical, falling towards zero as the distance between the two vectors gets large. A range of alternative radial basis functions exists including the inverse multi-quadratic function and the spline function.

The second phase of the model construction process is the determination of the value of the weights on the connections between the hidden layer and the output layer. In training these weights, the output value for each input vector will be known, as will the activation values for that input vector at each hidden layer node, so a supervised learning method can be used. The simplest transfer function for the node(s) in the output layer is a linear function where the network's output is a linearly weighted sum of the outputs from the hidden nodes. In this case, the weights on the arcs to the output node(s) can be found using linear regression, with the weight values being the regression coefficients. Sometimes it may be preferred to implement a non-linear transfer function at the output node(s). For example, when the RBFN is acting as a binary classifier it would be useful to use a sigmoid transfer function to limit outputs to the range $0 \rightarrow 1$. In this case, the weights between the hidden and output layer could be determined using the backpropagation algorithm.

Once the RBFN has been constructed using a training set of input-output data vectors it can then be used to classify or to predict outputs for new input data vectors, for which an output value is not known. The new input data vector is presented to the network, and an activation value is calculated for each hidden node. Assuming that a linear transfer function is used in the output node(s), the final output produced by the network is the weighted sum of the activation values from the hidden layer, where these weights are the coefficient values obtained in the linear regression step during training. The basic algorithm for the canonical RBFN is as follows:

i. Select the initial number of centres (m).
ii. Select the initial location of each of the centres in the data space.
iii. For each input data vector/centre pairing calculate the activation value $\phi(||x - y||)$, where ϕ is a radial basis function and $||...||$ is a distance measure between input vector x and a centre y in the data space. As an example, let $d = ||x - y||$. The value of a Gaussian RBF is then given by

$y = \exp(\frac{-d^2}{2\sigma^2})$, where σ is a modeller selected parameter which determines the size of the region of input space a given centre will respond to.

iv. Once all the activation values for each input vector have been obtained, calculate the weights for the connections between the hidden and output layers using linear regression.

v. Go to step (iii) and repeat until a stopping condition is reached.

vi. Improve the fit of the RBFN to the training data by adjusting some or all of the following: the number of centres, their location, or the width of the radial basis functions.

As the number of centres increases, the predictive ability of the RBFN on the training data will increase, possibly leading to overfit and poor out-of-sample generalisation. Hence, the object is to choose a sufficient number of hidden layer nodes to capture the essential features in the training data, without overfitting the training data. The selection of centre locations and the training of the RBFN can be automated by using an evolutionary algorithm (Chap. 3).

2.4 Self-organising Maps

Self-organising maps [128, 129, 130] are a form of NN which can cluster data using unsupervised learning. Unsupervised learning is used when the outputs (clusters) are not known a priori. This may occur for example, when trying to segment a customer base.

The SOM projects (compresses) input data vectors onto a low-dimensional space, typically a two-dimensional grid structure, thereby producing a visual representation of the input data. The unsupervised learning process is based solely on measures of similarity amongst the input data vectors. During the training process, the network undergoes *self-organisation* as 'like' input data patterns are grouped or clustered together on the grid. SOMs have been utilised for a variety of clustering and classification problems including speech recognition and medical diagnosis [101]. The SOM bears similarities with the traditional statistical technique of Principal Component Analysis (PCA). However, unlike PCA the projection of the input data is not restricted to being linear.

The SOM consists of two layers, the input layer (a holding point for the input data), and the *mapping* layer (Fig. 2.15). The input layer has as many nodes as there are input variables. The two layers are fully connected to each other and each of the nodes in the hidden layer has an associated weight vector, with one weight for each connection with the input layer.

The aim of the SOM is to group like input data vectors together on the mapping layer, therefore the method is *topology preserving* as items which are close in the input space are also close in the mapping space. During training the data vectors are presented to the SOM through the input layer one at a

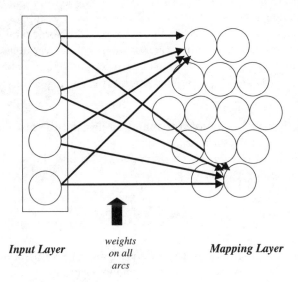

Fig. 2.15. A SOM with a 2-d mapping layer. On grounds of visual clarity, only the connections between the input layer and two of the mapping layer nodes are shown

time. The nodes in the mapping layer *compete* for the input data vector. The winner is the mapping node whose vector of incoming connection weights most closely resembles the components of the input data vector. The winner has the values of its weight vector adjusted to move them towards the values of the input data vector, and the mapping layer nodes in the neighbourhood of the winning node also have their weight vectors altered to become more like the input data vector (a form of *co-operation* between the neighbouring nodes). As more input data vectors are passed through the network, the weight vectors of the mapping layer nodes will self-organise. By the end of the training process, different parts of the mapping layer will respond strongly to specific regions of input space. Once training of the network is complete, the clusters obtained can be examined in order to gain better insight into the underlying dataset (for example, what input items have been grouped together, what are the typical values for each input in a specific cluster).

2.4.1 Implementing a SOM

A SOM can be implemented in a variety of ways. Specific choices faced by the modeller include the method of weight initialisation between the input and mapping layer nodes, the choice of topology of the mapping layer, the neighbourhood size, the distance measure employed in determining which mapping node is closest to a given input data vector, and the learning method used to update the weight vectors of the mapping layer once a winning node is determined. Input data vectors are typically standardised before training. Methods of standardisation include dividing each column of input variables by its standard deviation, or the standardisation of each column of inputs based on their range (e.g., $x^* = \frac{x - min(x)}{max(x) - min(x)}$). The general training algorithm for the SOM is as follows:

  i. Initialise the weights between the input nodes and the mapping nodes.
  ii. Present an input vector **x**: $x_0, x_1, ..., x_{n-1}$.
 iii. Calculate the distance between the input vector and the weight vector for each mapping layer node j

$$d_j = \sum_{i=0}^{n-1} (x_i - w_{ij})^2 \qquad (2.9)$$

 iv. Select the mapping node j^* that has the minimum value of d_j.
  v. Update the weight vector for mapping node j^* and its neighbouring mapping nodes as follows

$$w(t+1)_{ij} = w(t)_{ij} + \eta(t)h(t)(x_i - w_{ij}) \qquad (2.10)$$

where η is the learning rate of the map, and h defines a neighbourhood function. Both the neighbourhood size and the learning rate decay during the training run, in order to fine-tune the developing SOM.
 vi. Repeat steps (ii)-(v) until the weights have stabilised.

Classification with SOMs

Although technically SOMs are clustering algorithms, they can be used for classification purposes. For example, if the classification problem is binary and class membership is denoted by a 0 (class 1) or a 1 (class 2), and each class is considered equally likely to occur a priori, a ratio value approach can be applied. Under this approach, each node on the mapping layer outputs a value ($0 \le r \le 1$), where r is calculated using:

$$r = \begin{cases} 0.5 & \text{if } n_1 + n_2 = 0, \\ \frac{n_1}{n_1 + n_2} & \text{otherwise} \end{cases} \qquad (2.11)$$

where n_1 and n_2 are the number of class 1 and class 2 cases in the training set which have been mapped to that node. Values close to 0 indicate a node which classifies class 2 items, and values close to 1 indicate a node classifying class 1 items. Out-of-sample data vectors for which the classification is not known can be classified by presenting them to the network and determining the class label of the mapping node to which the input vector is closest.

If there is a known difference in the rate of occurrence of each class, a simple voting scheme can be used to assign class labels to each mapping node after training. Suppose for example that class 2 is known to occur more frequently than class 1. The default label for all mapping nodes is set as class 2. During the training process, mapping nodes which are found to respond more strongly to class 1 data vectors than to class 2 data vectors are relabelled as class 1. After training is completed, out-of-sample data vectors are classified by presenting them to the relabelled map, and assigning them the label of the mapping node they are closest to.

Once the initial assignment of class labels has been made, it is possible to fine-tune the mapping using learning vector quantisation (LVQ), a supervised learning methodology [129]. Under LVQ, the weight vectors for all mapping nodes are iteratively updated using

$$
\Delta w_i = \begin{cases} \eta(x - w_i) & \text{if } x \text{ is classified correctly,} \\ -\eta(x - w_i) & \text{if } x \text{ is classified incorrectly} \end{cases} \tag{2.12}
$$

where w_i is the weight vector for mapping node i. Typically a small value is set for η ($0.01 \rightarrow 0.02$), and it decrements to zero as the algorithm runs. The object of the fine-tuning step is to pull weight vectors of nodes in separate classes away from each other, in order to improve the delineation of the class boundaries on the map.

Financial Application of a Clustering Algorithm

Apart from their use in data-mining large databases for previously unknown patterns, clustering algorithms can be applied in the construction of a multi-stage trading system. One problem when developing trading systems is that the market is non-stationary, and different market conditions or *regimes* prevail from time to time. Therefore, a trading system which works well in one market environment may not perform well in another, and a practical difficulty emerges in deciding when to turn off or retrain an existing trading system. Unsupervised learning methodologies such as SOMs can be applied as a detector of a changing market environment. A SOM can be trained using historic market data, and thereafter used to assess whether current market data suggests that the market is moving from its current cluster or attractor to another. Detecting the beginning of such a movement may provide time to unwind trading positions and to exit from the current trading system before losses emerge.

2.5 Summary

NNs consist of a family of robust, data-driven modelling methodologies which generally outperform traditional linear modelling techniques when applied to financial data. However, the earlier comments regarding the clarity of NN models should be borne in mind. Traditional approaches have the virtue of apparent, if perhaps unwarranted, simplicity in terms of their model specification. A charge which is sometimes levelled against NN techniques is that they result in a *black box model* as it can be difficult to interpret their internal workings and understand why the model is producing its output. However, this criticism generally fails to consider that any truly complex, non-linear system is unlikely to be amenable to simple explanation. Despite the powerful modelling capabilities of NNs, they do suffer from a number of practical drawbacks:

   i. It is difficult to embed existing knowledge in the model, particularly non-quantitative knowledge.
  ii. Care must be taken to ensure that the developed models generalise beyond their training data.
 iii. Results from the commonly used MLP methodology are sensitive to the choice of initial connection weights.
  iv. The NN model-development process entails substantial modeller intervention, and can be time-consuming.

The last two of these concerns can be mitigated by melding the methodology with an evolutionary algorithm. The resulting hybrid models are discussed in the next chapter which introduces evolutionary algorithms.

3

Evolutionary Methodologies

This chapter provides an overview of a series of biologically inspired algorithms drawn from an evolutionary metaphor. A substantial literature exists which applies evolutionary methodologies to modelling of financial markets [2, 16, 51, 161].

In biological evolution, species are positively or negatively selected depending on their relative success in surviving and reproducing in their current environment. Differential survival and variety generation during reproduction provide the engine for evolution [49, 201]. These concepts have metaphorically inspired a family of algorithms known as evolutionary computation (EC). The chapter focusses on three evolutionary algorithms (EAs) that fall under the umbrella of EC: the genetic algorithm (GA), differential evolution (DE), and genetic programming (GP).

Members of the family of EAs share a great deal in common with each other and as such we shall treat the following overview of the concepts that underpin a specific EA instance, the GA, in the broader context of providing an introduction to some of their common principles.

3.1 Genetic Algorithm

Although the development of the GA dates from the 1960s, they were first brought to the attention of a wide audience by Holland [108]. To date their main application in a business setting has been in the domain of finance [16, 43, 51, 52, 217].

The GA is a mathematical optimisation algorithm with global search potential. The methodology is inspired by a biological metaphor and applies a pseudo-Darwinian process to *evolve* good solutions to real-world problems. The GA adopts a populational unit of analysis, wherein each member of the population encodes a potential solution to the problem of interest. These solutions may be as diverse as a set of rules, a series of coefficient values, or a representation of a NN. Evolution in the population of encodings is simulated by means of a pseudo-natural selection process using differential-fitness selection and pseudo-genetic operators which induce variation in the population between successive generations.

Although many variants of GAs exist [91, 153] each potential solution is traditionally encoded as a binary string (0101 . . . 1). The quality of each binary string is determined by reference to a problem-specific fitness function, which maps each string to a number representing its quality or fitness. The fitness of a string is typically, but not necessarily, normalised to the range 0 to 1 and standardised such that if 0 represents the worst possible fitness value then 1 corresponds to the best possible fitness, or vice versa. The fitness function represents an analogue of the environment in the real world. In addition to absolute measures of fitness as just described, it is also possible to define fitness relative to other members of the population without explicitly calculating values for each population member. For example, if our problem domain is to evolve a chess player we could evaluate the population by allowing individuals to play tournaments against each other where the winner of the tournament is deemed the fittest.

Just as biological genotypes encode the results of past evolutionary trials, the GA search heuristic encodes a history (or memory) as future populations are developed from the current population. Mathematically, the canonical GA can be formulated as a finite-dimension Markov chain, wherein each state corresponds to a configuration of the population of bit strings. The GA is a Markov process as the only memory that the GA has is that of the fitness of its population of strings in their last fitness evaluation. Depending on the form of genetic operators implemented in the algorithm, the transition probabilities between states will vary. In the canonical GA, the inclusion of a mutation operator implies that there are no absorbing states in the Markov process and that all states can potentially be visited.

Evolutionary algorithms, including the canonical GA, can be characterised as [79]:

$$x[t + 1] = r(v(s(x[t]))) \tag{3.1}$$

where $x[t]$ is the population of encodings at iteration t, $v(.)$ is the random variation operator (crossover and mutation), $s(.)$ is the selection for mating operator, and $r(.)$ is the replacement selection operator. Once the initial population of strings encoding solutions has been obtained and evaluated, a reproductive process is applied in which the encodings corresponding to the better-quality solutions have a higher chance of being selected for propagation of their genes into the next generation. In the canonical GA (with fitness-proportionate selection), the expected number of offspring for each encoding is given by $\frac{P_{obs}}{P_{ave}}$, where P_{obs} is the observed performance (fitness) of the corresponding solution and P_{ave} is the average performance of all solutions in the current population. Thus high-quality solution encodings may be chosen for replication several times in a single generation. Over a series of generations, the better adapted solutions in terms of the given fitness function tend to flourish, and the poorer solutions tend to disappear.

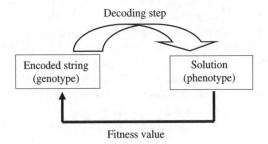

Fig. 3.1. Decoding of genotype into a solution in order to calculate fitness

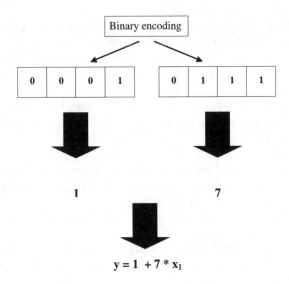

Fig. 3.2. Two-step decoding of a binary string into two integer values, which for example, represent the coefficients in a linear model

Therefore the canonical GA can be described as an algorithm that turns one population of candidate encodings and corresponding solutions into another using a number of stochastic operators. Selection exploits information in the current population, concentrating interest on high-fitness solutions. Crossover and mutation perturb these solutions in an attempt to uncover better solutions. Mutation does this by introducing new gene values into the population, while crossover allows the recombination of fragments of existing solutions to create new ones. It is important to note that the evolutionary process of a GA operates on the *encodings* of the solutions, rather than directly on the solutions themselves. In determining the fitness of these encodings, they must first be translated into a solution to the problem of interest, the fitness of this solution is determined, and finally this fitness is associated with the encoding (Fig. 3.1). Figure 3.2 demonstrates a sample decoding of a binary string to produce the coefficients for a linear model.

3.1.1 Canonical GA

To provide an overview of the operation of the canonical GA, a flowchart (Fig. 3.3) and description of the steps in the algorithm is provided. The key steps in the algorithm are:

i. Determine how the solution is to be encoded as a string, and determine the definition of the fitness function.
ii. Construct an initial population, possibly randomly, of n encodings corresponding to candidate solutions to a problem.
iii. Decode each string into a solution, and calculate the fitness of each solution candidate in the population.
iv. Implement a selection process to select a pair of encodings corresponding to candidate solutions (the *parents*) from the existing population, biasing the selection process in favour of the encodings corresponding to better/fitter solutions.
v. With a probability p_{cross}, perform a crossover process on the encodings of the selected parent solutions, to produce two new (*child*) solutions.
vi. Apply a mutation process, with probability p_{mut}, to each element of the encodings of the two child solutions.
vii. Store the encodings corresponding to the child solutions in the new (next generation) population.
viii. Repeat steps (iv)-(vii) until n encodings of candidate solutions have been created in the new population. Then discard the old population. This constitutes a generation.
ix. Go to step (iii) and repeat until the desired population fitness level has been reached or until a predetermined number of generations have elapsed.

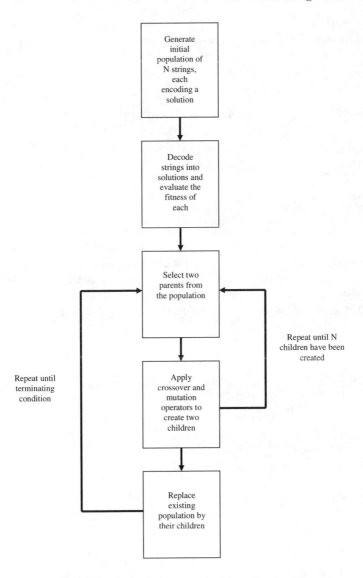

Fig. 3.3. Flowchart of the canonical genetic algorithm

3.1.2 Example of the GA

To provide additional insight into the workings of the canonical GA, a numerical example is now provided. Assume that candidate solutions are encoded as a binary string of length 8 and the fitness function $f(x)$ is defined as the number of ones in the bit string (this is known as the OneMax problem). Let

$n = 4$ with $p_{cross} = 0.7$ and $p_{mut} = 0.001$. Assume also that the initially generated (random) and evaluated population is that in Table 3.1.

Table 3.1. An example initial random population

Candidate	String	Fitness
A	00000110	2
B	11101110	6
C	00100000	1
D	00110100	3

Next a selection process is applied based on the fitness of the candidate solutions. Suppose the first selection draws candidates B and C and the second draws B and D. For each set of parents, the probability that a crossover (recombination) operator is applied is p_{cross}. Assume that B and C are crossed over between bit position one and two (arbitrary) to produce child candidates E and F (Table 3.2), and that crossover is not applied to B and D.

Table 3.2. Crossover applied to individuals B and C from Table 3.1, after the first element, to produce the offspring E and F

Initial Parent	**Candidate B**	**Candidate C**
	1 1101110	0 0100000
Resulting Child	**Candidate E**	**Candidate F**
	0 1101110	1 0100000

Crossover is not applied to B and D, hence the child candidates (G and H) are clones of the two parent candidates (Table 3.3).

Table 3.3. No crossover is applied to B and D, hence the child candidates G and H are clones of their parents

Initial Parent	**Candidate B**	**Candidate D**
	11101110	00110100
Resulting Child	**Candidate G**	**Candidate H**
	11101110	00110100

Finally, the mutation operator is applied to each child candidate with probability p_{mut}. Suppose candidate E is mutated (to a 0) at the third locus and that no other mutations take place. The resulting new population is presented in Table 3.4.

By biasing selection for reproduction towards more fit parents, the GA has increased the average fitness of the population in this example from 3 ($\frac{2+6+1+3}{4}$) to 3.75 ($\frac{4+4+6+3}{4}$) after the first generation.

Table 3.4. Final new generation of solutions after mutation operator has been applied

Candidate	String	Fitness
E	01001110	4
F	10100000	2
G	11101110	6
H	00110100	3

3.1.3 Extending the Canonical GA

The two subsections above describe and provide a simple worked example of a canonical GA. A large number of variants on this basic algorithm exist, all of which fall under the GA umbrella. This subsection briefly describes a number of these variants.

Genotype Encoding

In assessing the utility of a specific genotype, it must first be decoded into its associated phenotype. Although binary encodings are often used in GAs, there are multiple ways that binary strings can be decoded to produce integer or real values. The simplest decoding method is to convert the binary string to an integer value, which can in turn be converted into a real value if required. A binary genotype of length n can encode any integer from 0 to $2^n - 1$ (Table 3.5). If a real-valued output is required, the integer value obtained by decoding the binary string can be divided by $2^n - 1$ to obtain a real number in the interval [0,1]. A real number in any interval $a \rightarrow b$ can be obtained by taking the result of the last calculation and rescaling it using the formula $a + x(b - a)$.

Taking an example, a binary string which is eight bits long can encode any integer between 0 and 255. If we consider the binary string (00000111), this can be decoded into the integer value 7 (calculated as: $2^0 \times 1 + 2^1 \times 1 + 2^2 \times 1 + 2^3 \times 0 + 2^4 \times 0 + 2^5 \times 0 + 2^6 \times 0 + 2^7 \times 0$). If instead of an integer value in the range 0 to 255, a real value in the range $0 \rightarrow 5$ was required, the integer value could be converted into a real value as follows: $0 + \frac{7}{255} \times (5 - 0) = 0.027451$.

Although the above decoding scheme for a binary string is quite simple, it can suffer from Hamming cliffs, in that sometimes a large change in the genotype is required to produce a small change in the resulting integer value. Looking at the change in the binary value required to move from an integer value of 3 to 4 in Table 3.5, it can be seen that the underlying genotype needs to change in all three bit positions. These Hamming cliffs can potentially create barriers that the GA could find difficulty in passing. In contrast, other schemes such as Gray coding reduce this problem. In Gray coding, the object is to create a code such that a single integer change only requires a 1-bit change in the binary genotype. This means that adjacent solutions in the (integer or real-valued) search space will be adjacent in the (binary) encoding space as well, requiring fewer mutations to discover. The Gray coding rule starts with a string of all zeros for the integer value zero, and to create each subsequent integer in sequence the rule successively flips the right-most bit that produces a new string. Despite the apparent potential benefits of Gray coding, it will not necessarily produce better results than the canonical binary coding system.

Table 3.5. Integer conversion for standard and Gray coding

Integer Value	Canonical Binary Code	Gray Code
0	000	000
1	001	001
2	010	011
3	011	010
4	100	110
5	101	111
6	110	101
7	111	100

For some problems, a real-valued encoding is a more natural representation than a binary encoding. This raises the question as to what modifications should be made to the mutation and the crossover operators for the real-valued case. A simple strategy for modifying mutation is to implement a stochastic mutation operator, where an element of a real-valued string can be mutated by adding a small (positive or negative) real value to it. The crossover operator can be modified so that elements from the string of each parent are averaged in order to produce the corresponding value in their child. Many mutation and crossover schemes for real-valued encodings exist.

Measuring Fitness

When applying the GA, an objective function is required for the problem of interest, and the objective function value for each phenotype is transformed into a fitness measure using a fitness function. Thus

$$F(x) = g(f(x))$$

where f is the objective function, g transforms the value of the objective function into a non-negative number, and F is the fitness measure. If the value of the objective function is always non-negative, the raw objective function values can be rescaled using a linear equation of the form:

$$g_1 = af_1 + b$$

where a is chosen in order to ensure that the maximum fitness value is a scaled multiple of the average fitness and b is chosen in order to ensure that the resulting fitness values are non-negative. The reason for using rescaled fitnesses rather than raw objective function values is to control the selection pressure in the algorithm (see next subsection). However, simple linear scaling can still result in high selection pressure and rapid convergence of the population. Alternative transformation methods exist such as sigma scaling and Boltzmann selection.

Selection Methods

A key issue in designing a good GA for a specific problem is the management of the *exploration vs exploitation* trade-off. The algorithm must utilise, or exploit, already discovered fit solution encodings, while not neglecting to continue to explore new regions of the search space which may contain yet-fitter solution encodings. Choices for the selection process and the diversity generating operators of mutation and recombination determine the balance between exploration and exploitation.

The original and the simplest method of selection for reproduction in the GA is fitness-proportionate selection, whereby the probability that a specific member of the current population is selected for mating is directly related to its relative fitness. The selection process is therefore directed towards good members of the current population.

A problem with this method of selection is that it can lead to premature convergence of the population of encodings. Fitness-proportionate selection embeds a high *selection-pressure*, and can force too much selection of high-fitness individuals. Commonly, in the early stage of the search process there is a high variance in the fitness of solution encodings, with a small number of encodings being notably fitter than the others. These encodings and their descendants can overrun the entire population, thereby reducing the subsequent exploration of the search space. Better selection schemes will encourage exploitation of high-fitness individuals in the population, without losing diversity in the population too quickly.

Several alternative selection methods have been designed to overcome the problems of fitness-proportionate selection. One computationally efficient method is that of tournament selection. Under tournament selection, k members are chosen randomly without replacement from the population, and the fittest of these is chosen as the tournament winner and is 'selected' to act as a parent. Assuming a population of size N, the value of k can be varied within $2, \ldots, N$. Lower values of k provide lower selection pressure, while higher values provide higher selection pressure. For example, if $k = N$, the fittest individual is always the tournament winner.

Crossover and Mutation

The mutation operator is important in the GA as it ensures that the search process never stops. At each iteration mutation can potentially uncover useful novelty. In contrast, crossover ceases to generate novelty once all members of the population converge to a single string form. The rate of mutation also has important implications for the utility of selection and crossover. If a very high rate of mutation is applied, the selection and crossover operators can be overpowered, and the GA will more closely resemble random search. Hence the aim is to select a rate of mutation which helps generate useful novelty, but which does not rapidly destroy good solutions which are being exploited through selection and crossover. Of course, there is no requirement that the mutation rate must remain constant during the GA run. One strategy in designing a GA could be to vary the rate of mutation depending on the degree of similarity in the population of strings. If the population is converging, the rate of mutation can be increased, in order to promote greater diversity in the population.

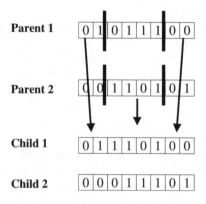

Fig. 3.4. Two-point crossover

Crossover serves two purposes in the GA. First, it allows for the inheritance of 'good genes' or partial solution encodings by the offspring of parents. It also serves to reduce the search space to regions of greater promise. Crossover can

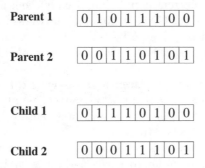

Parent 1 | 0 | 1 | 0 | 1 | 1 | 1 | 0 | 0 |

Parent 2 | 0 | 0 | 1 | 1 | 0 | 1 | 0 | 1 |

Child 1 | 0 | 1 | 1 | 1 | 0 | 1 | 0 | 0 |

Child 2 | 0 | 0 | 0 | 1 | 1 | 1 | 0 | 1 |

Fig. 3.5. Uniform crossover, where a random choice is made as to which parent donates a bit to child 1. Child 2 is then constructed using the bits not selected for inclusion in child 1

be implemented in a large number of ways. One problem of single (one) point crossover, is that related components of a solution encoding (schema) which are widely separated on the string tend to be disrupted when crossover is applied. One way of reducing this potential problem is to implement two-point crossover (Fig. 3.4), whereby two positions on the parent strings are chosen randomly and the segments between the two positions are exchanged.

A popular form of crossover is *uniform crossover*. One way of implementing uniform crossover is to compare the two parent strings element by element. In producing a child string, a random selection is made from each parent when filling each corresponding locus on the child's genotype. The process can be repeated a second time to create a second child, or the second child could be created using the values not selected when producing the first child (Fig. 3.5).

Another design strategy that is sometimes employed with GAs, particularly if the fitness landscape is likely to be multi-modal, is to use more than one population. In this implementation, known as the *island model*, several separate populations are created and commence their own evolutionary process. Periodically, fit individuals are allowed to 'migrate' between the sub-populations. The migrations promote the sharing of the information in good solution encodings, while maintaining genotypic diversity between the sub-populations.

Replacement Strategies

In the canonical GA, a generational replacement strategy was adopted, whereby the entire current population was replaced by the newly created population of child encodings. Alternative child generation and replacement strategies include *steady state* replacement, where only a small number of children (sometimes only one) are created during each generation, with only a small number of the current population, usually the least fit, being replaced during each iteration of the GA. For example, the worst x members of the current population could be replaced by the best x children. Adopting a steady state replacement strategy ensures that successive populations overlap to a significant degree.

Another commonly used replacement strategy is *elitism* whereby the best member (or several best members) of the current population always survive into the next population. This strategy ensures that a good individual is not lost between successive generations.

Some GA applications use *crowding operators* as a supplement in their replacement strategy. In order to encourage diversity in the population of solution encodings, a new child solution is only allowed to enter the population by replacing the current member of the population which is most similar to itself. The objective is to avoid having too many similar individuals (crowds) in the population.

While the above exposition outlines the primary components and principles upon which the GA is based it is by no means an exhaustive overview of GAs. For descriptions of additional genetic operators, selection operators, and advances in GA theory the reader is encouraged to explore the extensive literature in this area. A sample of notable developments in this field is outlined in Sect. 3.6.

3.1.4 Schema and Building Blocks

The computational power of GAs results from their *explicit* and *implicit* parallel processing capabilities. The explicit parallelisation stems from their maintenance of an entire population of data vectors, rather than a single data vector. The implicit parallel processing capabilities arise due to the *Schema Theorem* [108] which demonstrates that under general conditions, in the presence of differential selection, crossover and mutation, almost any compact cluster of components (bits) that provides above-average fitness will grow exponentially in the population from one generation to the next. Schema are analogous to templates for different bit combinations. For example in a binary representation, the schema 1**1 represents all four bit strings which begin and end with a 1. The symbol *, commonly referred to as a *don't care symbol*, is used as a placeholder for either a 0 or a 1, such that it is irrelevant which of these symbols appears at these positions of a schema. Two important characteristics of schema are their defining length and their order. The defining length of a schema is the distance between the two furthest-apart components of the schema which are not don't care symbols. The order of a schema is the number of components which are not 'don't care' symbols. For example, the order of the above schema is two and its defining length is four. Further examples are provided in Table 3.6.

The parallel nature of a GA search process makes it less vulnerable to local optima than traditional hill climbing optimisation methods. Despite the

Table 3.6. Example schemata with their order and defining length provided

Schema	Defining Length	Order
*0*1	3	2
**1*	1	1
1111	4	4
****	0	0

good properties of GAs they, like all non-linear optimisation techniques, are subject to limitations. The methodology is optimising but there is no guarantee that an optimal solution will be found in finite time. Progress towards better solutions may be intermittent rather than gradual. Consequently, the time required to find a high-quality solution to a problem is not determinable ex ante.

3.2 Differential Evolution

Differential evolution (DE) [181, 204, 205, 206] is a population-based search algorithm. The algorithm draws inspiration from the field of evolutionary computation, as it embeds implicit concepts of mutation, recombination and fitness-based selection to evolve good solutions to a problem of interest by manipulating a population of solution encodings. It also borrows principles from social algorithms (Chps. 5 and 6) in the manner in which new individuals are generated. Unlike the binary chromosomes typical of canonical GAs, an individual in DE is generally comprised of a real-valued chromosome.

3.2.1 DE Algorithm

Although several DE algorithms exist, we primarily describe one version of the algorithm based on the *DE/rand/1/bin* scheme [204]. The different variants of the DE algorithm are described using the shorthand $DE/x/y/z$, where x specifies how the base vector (of real values) is chosen (*rand* if it is randomly selected, or *best* if the best individual in the population is selected), y is the number of difference vectors used, and z denotes the crossover scheme (*bin* for crossover based on independent binomial experiments, and *exp* for exponential crossover).

At the start of the algorithm, a population of N, d-dimensional vectors $X_j = (x_{i1}, x_{i2}, \ldots, x_{id})$, $j = 1, \ldots, N$, each of which encode a solution, is randomly initialised and evaluated using a fitness function f. During the search process, each individual (j) is iteratively refined. The modification process has three steps:

i. Create a *variant vector* which encodes a solution, using randomly selected members of the population (mutation step).

ii. Create a *trial vector*, by combining the variant vector with j (crossover step).

iii. Perform a *selection process* to determine whether the newly-created trial vector replaces j in the population.

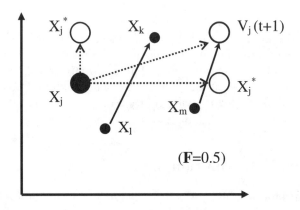

Fig. 3.6. A representation of the differential evolution variety-generation process. The value of F is set at 0.50. In a simple 2-d case, the child of particle X_j can end up in any of three positions. It may end up at either of the two positions X_j^*, or at the position of particle $V_j(t+1)$

Under the mutation operator, for each vector $X_j(t)$ a variant vector $V_j(t+1)$ is obtained:

$$V_j(t+1) = X_m(t) + F(X_k(t) - X_l(t)) \tag{3.2}$$

where $k, l, m \in 1, \ldots, N$ are mutually distinct, randomly selected indices, and all the indices $\neq j$ (X_m is referred to as the base vector, and $X_k(t) - X_l(t)$ is referred to as a difference vector). Selecting the three indices randomly implies that all members of the current population have the same chance of being selected, and therefore influencing the creation of the difference vector. The difference between vectors X_k and X_l is multiplied by a scaling parameter F (typically $F \in (0, 2]$). The scaling factor controls the amplification of the difference between X_k and X_l, and is used to avoid stagnation of the search process.

There are several alternative versions of the above process for creating a variant vector [204]. It is possible to use more than one difference vector. For example, five indices could be randomly selected (*DE/rand/2/bin*), and the variant vector calculated:

$$V_j(t+1) = X_m(t) + F(X_k(t) - X_l(t)) + F(X_o(t) - X_p(t)) \qquad (3.3)$$

Another alternative is to include the highest-fitness member of the current population when calculating the variant vector ($DE/best/1/bin$), for example:

$$V_j(t+1) = X_{best}(t) + F(X_k(t) - X_l(t)) \qquad (3.4)$$

This bears similarity with the use of *gbest* in the particle swarm algorithm (Chap. 5), as the current best member of the population has an impact on the generation of all trial vectors. This implicitly increases the selection pressure in the algorithm. Many alternative methods for selecting a base vector could be employed including for example, tournament selection.

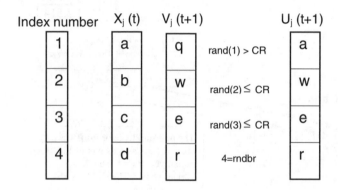

Fig. 3.7. An example of crossover in DE

A notable attribute of the mutation step in DE is that it is self-scaling. The size/rate of mutation along each dimension stems solely from the location of the particles in the current population. The mutation step self-adapts as the population converges leading to a finer-grained search. In contrast, the mutation process in the canonical GA is typically based on draws from a separately defined (fixed) probability density function.

Following the creation of the variant vector, a trial vector $U_j(t + 1) = (u_{j1}, u_{j2}, \ldots, u_{jd})$ is obtained:

$$U_{jk}(t+1) = \begin{cases} V_{jk}(t+1), & \text{if } (rand \le CR) \text{ or } (j = rnbr(ind)) ; \\ X_{jk}(t), & \text{if } (rand > CR) \text{ and } (j \ne rnbr(ind)). \end{cases} \qquad (3.5)$$

where $k = 1, 2, \ldots, d$, *rand* is a random number generated in the range (0,1), CR is the user-specified crossover constant from the range (0,1), and $rnbr(ind)$

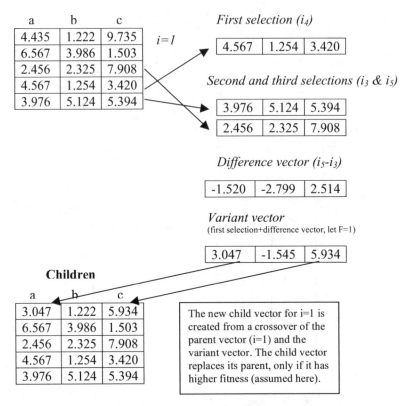

Fig. 3.8. Numerical example of the canonical DE algorithm

is a randomly chosen index chosen from the range $(1, 2, \ldots, d)$. The random index is used to ensure that the trial solution differs by at least one component from $X_j(t)$.

The resulting trial (child) solution replaces its parent if it has higher fitness (a form of selection), otherwise the parent survives unchanged into the next iteration of the algorithm (Fig. 3.6). Figure 3.7 provides an illustration of the crossover operator in DE, and Fig. 3.8 illustrates a simple numerical example. In the numerical example, the parent vector is i=1. Three other vectors are randomly chosen to create the variant vector, and F=1 is assumed. When crossover is applied between the parent and the variant vector, the first and the third elements of the variant vector are assumed to combine with the second element of the parent vector to create the trial or child vector. Finally, it is assumed that the fitness of the trial vector exceeds that of its parent and it therefore replaces the parent.

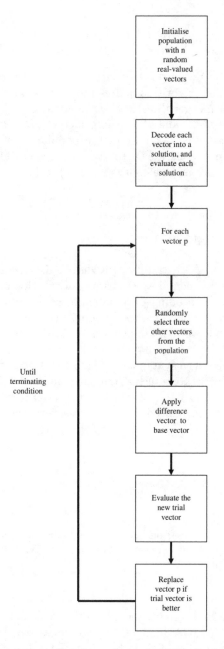

Fig. 3.9. A flowchart of a typical DE algorithm

$$X_j(t+1) = \begin{cases} U_j(t+1), & \text{if } f(U_j(t+1)) > f(X_j(t)); \\ X_j(t), & \text{otherwise.} \end{cases} \quad (3.6)$$

Figure 3.6 provides a graphic of the adaptive process described above, and an outline of a typical DE algorithm is presented in Fig. 3.9.

The DE algorithm has three key parameters: the population size (N), the crossover rate (CR), and the scaling factor (F). Higher values of CR tend to produce faster convergence of the population of solutions. Typical values for these parameters are in the ranges, N=50-100 (or five-ten times the number of dimensions in a solution vector), CR=0.4-0.7 and F=0.4-0.9 for the DE/rand/bin scheme.

3.3 Genetic Programming

Genetic programming (GP) traditionally distinguishes itself from the genetic algorithm in two fundamental ways. Instead of evolving binary strings which represent an indirect encoding of a potential solution, in GP evolutionary search is applied to the solution directly, solutions in this case being computer programs (or, alternatively, trading systems). In the form of GP popularised by John Koza [131, 132, 133, 134] these take the form of Lisp S-expressions, which are represented as a syntax tree (Fig. 3.10). It is to these trees that the evolutionary search operators such as crossover and mutation are applied. Figure 3.11 illustrates an example of two parent syntax trees, and demonstrates how sub-trees can be exchanged during a crossover event.

(+ (sin x) (* x 3.14) (/ y x))

Fig. 3.10. Example S-expression (left) and corresponding syntax tree (right). The syntax tree decodes into the expression (sin x) + (3.14x) + (y/x), where x and y are predefined constants

The second fundamental difference is in the variable-length representation adopted by GP. With GA (and in DE) we adopt a fixed-length encoding, whereby we fix the number of genes (or bits) that will comprise an individual at the outset of a run. In GP it is recognised that the length of a solution program may not be known a priori and as such the number of genes must itself be open to evolution. Initialisation of a GP population consequently attempts to generate a diversity not only in the values of the genes (the

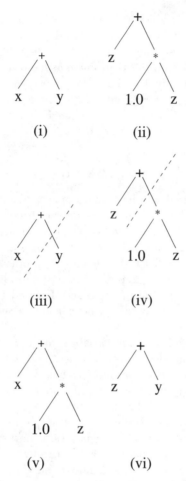

Fig. 3.11. Example syntax trees, (i) and (ii), undergoing a crossover event to produce two child trees, (v) and (vi). The hashed lines on syntax trees (iii) and (iv) represent the site at which crossover takes place

primitive symbols of the programming language) but also in the structure of the individuals.

Function and Terminal Sets

When evolving programs there are a number of issues that must be taken into consideration. The programs are generated using elements from two sets, namely, the function and terminal sets. The function set contains functions that have an arity greater than zero, while the terminal set contains functions that have an arity of zero, the arity of a function referring to the number of arguments it can take. For example, the functions *sin* and *not* have an arity of

1, while constants and variables have an arity of zero as they do not take any arguments. The members of the function and terminal sets must be chosen such that together they are *sufficient*, that is, they are powerful enough to represent a complete solution to the problem at hand. It must also be ensured that the function and terminal set have the property of *closure*. That is, each function should handle gracefully all values it might ever receive as inputs. For example, a suitable function (denoted by F) and terminal set (T) for a boolean problem with three input variables is given below.

$$F = \{ \texttt{and, not} \}$$
$$T = \{ \texttt{input0, input1, input2} \}$$

The function and terminal sets hold the property of sufficiency, as it can be shown that all possible boolean functions on the three input variables can be constructed from the boolean **and** and boolean **not** operators alone. Similarly, boolean function sets { **or, not** }, { **nand** } or { **nor** } are possible alternatives to meet the property of sufficiency. The closure property is met because the boolean input values (**input0, input1, input2**) can all be passed as inputs to each of the functions in the function set (F), and the output from each function in F is also a boolean value that can be passed in turn as input to another function from this set.

As well as including the input variables for the problem at hand, the terminal set typically includes constant values. The standard approach to the provision of constants in GP is through *ephemeral random constants* (ERCs). A number of ERCs are generated in the initial population within a prespecified range at the outset of a run of the GP algorithm. When a node in the growing program is determined to have become a constant, a random value in the ERC range is generated. After the initial generation, new constants are created through the recombination of existing ERCs through arithmetic expressions. It is also possible to target constants within an individual for mutation usually within a prespecified deviation.

Initialisation Strategy

Once the function and terminal sets are specified, individuals in the population must be generated using an initialisation strategy. In order to ensure diversity of both structure and values it is common to adopt the *ramped-half-and-half* initialisation. Ramped-half-and-half combines the *Grow* and *Full* initialisation strategies, each of which is used to generate half of the population (Fig. 3.12). In the Full method, trees are grown randomly such that all branches reach a predetermined maximum node depth, while in the Grow method trees are grown randomly with no one branch allowed to exceed the maximum node depth. In addition to adopting sub-tree crossover, as outlined earlier, amongst other genetic operators variety can be created through sub-tree mutation. Sub-tree mutation involves randomly picking a sub-tree, deleting that sub-tree, and then growing a new sub-tree in a random manner similar to the initialisation process.

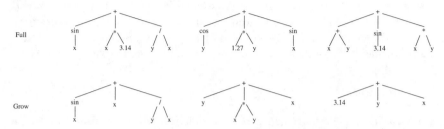

Fig. 3.12. Example GP population of size 6 created using the ramped-half-and-half initialisation strategy up to a tree depth of 3

In addition to the input variables constants, and primitive operators specified in the function and terminal sets, it is possible to incorporate standard programming constructs such as conditional statements, parameterised functions, iterations, loops, storage/memory, and recursion into a GP individual. An example GP program containing a conditional expression in both a prefix Lisp-like S-expression and syntax tree can be seen in Fig. 3.13. Note that the conditional expression, denoted by the `if` function at the root of the sub-tree, is comprised of three components. The first, left-most component is the condition itself, which can be comprised of a complex logical expression which will return either one of the boolean `true` or `false` values. In this example, depending on the outcome of the logical expression returning either true or false, the second (the value of x) or third component (value of y) of the conditional expression will be returned, respectively, and subsequently added to the result of `sin(x)`.

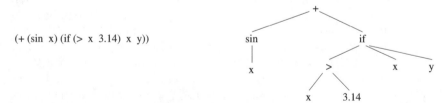

(+ (sin x) (if (> x 3.14) x y))

Fig. 3.13. Example GP individual containing a conditional S-expression (left) and its corresponding syntax tree (right)

3.3.1 More Complex GP Architectures

In the example GP individuals we have met so far, the programs are comprised of a single, result-producing function comprised of the whole tree. A wide range of more sophisticated GP architectures exist, in which the syntax tree can embed commonly found programming structures and concepts, including functions, memory, looping and recursion. Each of these is discussed below.

Functions in GP

In programming, and more generally in problem solving, it is useful to decompose the task at hand into a series of smaller and simpler sub-tasks, which can be reused to solve the problem as a whole. The ability to reuse parts of solutions can be incorporated into GP individuals using constructs such as functions, iterations, loops and recursion. To this end it is necessary to introduce a more complex program architecture comprised of multiple branches including the result-producing branch (RPB). The other branches define, for example, the functions and iterations that the RPB can utilise in the generation of the resulting program output.

The typical method to include functions or sub-routines in a GP individual is through *automatically defined functions* (ADFs). ADFs are parameterised functions that can be called in a hierarchical manner, and are defined in a function-defining branch comprised of the function's name, the list of its parameters, and the body of the function. Figure 3.14 outlines the architecture of a GP individual comprised of a single ADF called ADF0 that receives three parameters and sums the parameter values that are passed to it. An ADF is defined using the DEFUN function, and VALUES is a function that returns whatever value its sub-tree evaluates to. DEFUN simply returns the name of the function to its parent function (PROGN). The PROGN function evaluates all of its sub-trees in succession returning the result of evaluating the last (the right-most) sub-tree, which is referred to as the RPB. The RPB can use any of the previously defined ADFs when evaluating the result of the program.

An ADF may non-recursively call any previously defined ADF from within its own body, thus allowing hierarchical ADF evaluation. A succession of ADFs can thus precede the main RPB in an individual, as outlined in Fig. 3.15.

To ensure that architecturally correct (i.e., only permit non-recursive and hierarchical ADF calls to previously defined ADFs) individuals are generated in the initial population separate function and terminal sets must be specified for the ADFs and RPBs. Taking an example we could define two ADFs (ADF0 and ADF1), with the following function sets for the RPB, ADF0 and ADF1, respectively.

$$F_{RPB} = \{ \text{ if, } *, +, -, /, \text{ ADF0, ADF1 } \}$$
$$F_{ADF0} = \{ \text{ if, } *, +, -, / \}$$

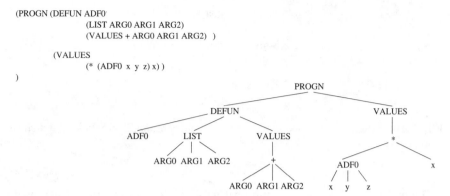

```
(PROGN (DEFUN ADF0
            (LIST ARG0 ARG1 ARG2)
            (VALUES + ARG0 ARG1 ARG2)  )

      (VALUES
            (* (ADF0 x y z) x) )
)
```

Fig. 3.14. The architecture of an automatically defined function represented in terms of an S-expression (top) and corresponding syntax tree (bottom). The syntax tree decodes to $(x+y+z)*x$

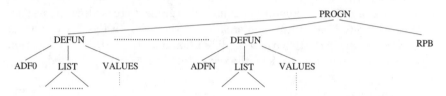

Fig. 3.15. The architecture of a GP individual including a hierarchy of automatically defined functions (ADF0 to ADFN) and the result-producing branch (RPB) represented as a syntax tree

$$F_{ADF1} = \{ \text{ if, } *, +, -, /, \text{ ADF0 } \}$$

The corresponding terminal sets for a problem with three variables might take the following form where ADF0 is a three-argument function and ADF1 has two arguments.

$$T_{RPB} = \{ \text{ x, y, z } \}$$
$$T_{ADF0} = \{ \text{ ARG0, ARG1, ARG2 } \}$$
$$T_{ADF1} = \{ \text{ ARG0, ARG1 } \}$$

A particular advantage of using ADFs when applying GP to design a trading system is that they allow for the easy (multiple) reuse of already discovered good code modules.

Memory in GP

Memory is implemented in GP in a manner similar to ADFs, using automatically defined storage (ADS), with the addition of two branches to an

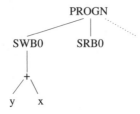

Fig. 3.16. Fragment of an example GP individual containing automatically defined storage (ADS0)

individual that allow reading and writing to a memory location. Effectively, the additions of a storage writing branch (SWB) and a storage reading branch (SRB) are equivalent to adding a new element to an individual's function set which allows a newly added memory location to be written to as well as read from (Fig. 3.16). The type (e.g., named memory, stack, queue, 2-dimensional array, or list) and dimensionality (number of arguments to address it) are determined (usually randomly) upon creation of the ADS. In Fig. 3.16 a named memory location (ADS0) with zero dimensionality (i.e., the SRB function requires no arguments to retrieve the data stored in ADS0) is created.

Looping in GP

Iterations and more generally loops can be incorporated into a GP individual using automatically defined iterations (ADIs) and automatically defined loops (ADLs). Similar to ADFs, ADIs and ADLs are defined using a multiple branch architecture, where their branches occur before the RPB. It is common for a simplified form of ADIs and ADLs to be adopted where the defined iterations or loops are invoked only once and prior to the evaluation of the RPB. The result of evaluating the ADI/ADL branch is made available to the RPB indirectly through storage in a named memory location. There may be multiple ADIs and ADLs within an individual, and they can refer to previously defined ADFs.

In the case of ADIs, they are implemented to iterate once over a predefined data structure such as an array, vector or sequence. As such, the size of the data structure to iterate over is known and the possibility for infinite loops to arise is eliminated. An example ADI is given in Fig. 3.17 which has no arguments, and returns the result of its evaluation indirectly to the result-producing branch by writing to the named memory location M0. The number of elements contained in the data structure being iterated over (V) is built into the ADI function. ADI0 is evaluated as a result of its invocation in the result-producing branch, with the RPB using the result of evaluating ADI0 by accessing M0.

ADLs implement a general form of iteration comprised of loop initialisation (LIB), loop condition (LCB), loop body (LBB), and loop update branches

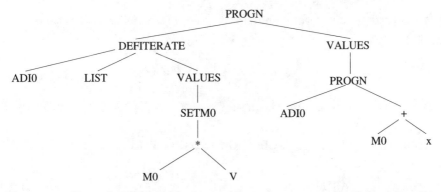

Fig. 3.17. Example GP individual containing an automatically defined iteration (ADI). The result of evaluating ADI0 (multiplying all the values contained in the vector V) is available to the result-producing branch through the named variable memory location M0, which the body of ADI0 wrote to using SETM0

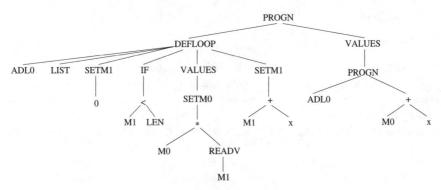

Fig. 3.18. Example GP individual containing an automatically defined loop (ADL). As in the ADI example in Fig. 3.17 the result of evaluating ADL0 is available to the result-producing branch through the named memory location M0, which the body of ADL0 wrote to using SETM0

(LUB). Figure 3.18 outlines an example ADL where the LIB sets the memory location M1 to 0; the LCB determines how many iterations over the data structure should be conducted. After the LBB is evaluated on each iteration the LUB is evaluated, which increments the value of M1 (in this example this ensures that an infinite loop will not arise as the LCB is checking the value of M1 to determine when to terminate the loop). In the LBB, (READV M1) reads the $M1^{th}$ value of the data structure (V) being looped over. The result of evaluating ADL0 is available to the RPB through the value stored in the named memory location M0. To prevent infinite loops occuring, generally a timeout strategy is adopted whereby the evaluation of an individual is halted after a predetermined time limit (or a maximum number of iterations) has been reached.

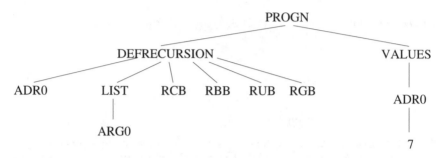

Fig. 3.19. The architecture of automatically defined recursion (ADR)

Recursion in GP

Recursion is made possible in GP through an automatically defined recursion (ADR) architecture (Fig. 3.19). There are four components to ADRs, namely recursion condition (RCB), recursion body (RBB), recursion update (RUB), and recursion ground (RGB) branches. To prevent infinite recursion a limit is placed on the number (or depth) of the recursive calls that are allowed within an individual. When timeout limits in the case of ADIs and ADLs, or depth limits in the case of ADRs are violated the individual can be selected against by punishment with a large fitness penalty. Figure 3.19 outlines ADR in an individual where the DEFRECURSION function is used to define the recursive function ADR0 that takes a single parameter (ARG0). The RCB determines if recursion is continued by returning a positive value, or in the case of returning a negative value recursion is halted. In the event recursion is halted the fourth (right-most) RGB branch is evaluated. The RBB branch normally contains a recursive call to the function itself, and when the evaluation of the RBB finishes the RUB branch is evaluated.

During initialisation of a GP population either the architecture of the program is prespecified, that is the presence (or absence) of ADFs, ADSs, ADIs/ADLs and ADRs and their quantities are predetermined, or their incorporation (deletion) can be left open to evolutionary search. In order to allow the search process to add, delete or modify these constructs architecture-altering operations were introduced specifically for each architecture type [133]. For example, in the case of ADFs, it is possible to create, duplicate or delete an ADF, and even to create, duplicate or delete arguments to an existing ADF. Special attention must also be paid to the crossover operator, which must be implemented to ensure that legal architectures are generated as a consequence of a crossover event.

A great deal of literature exists on genetic programming and the evolution of programs in general. The interested reader is referred to the following as a good starting point for further investigations [15, 131, 132, 133, 134, 136, 137, 138]. In Sect. 3.6 some of the more recent developments in genetic programming are discussed.

3.4 Combining EA and MLP Methodologies

Biologically inspired algorithms need not be applied on a stand-alone basis. They can also be combined in order to benefit from their complementary strengths. As an example of this, we will discuss a hybrid GA-MLP methodology, but it should be noted that a DE-MLP, or even GP-MLP, could also be used.

Despite the apparent dissimilarities between GAs and MLPs, the methods can usefully complement each other by combining the non-linear mapping capabilities of a MLP with the optimising capabilities of a GA. The construction of a MLP entails the selection of model inputs and model structure from many alternatives, and represents a combinatorial problem. An evolutionary algorithm such as the GA provides scope to automate this step.

There are several ways that GAs can be combined with a MLP, depending on what the modeller wishes to achieve. The first possibility is to use the GA to uncover a subset of good-quality model inputs from a possibly large set of potential inputs. In a time-series model, this could correspond to uncovering good-quality lag periods. A second use of GAs is to evolve the structure of a MLP network. Third, a GA could be used to evolve the choice of learning algorithm and relevant parameters for that algorithm. Therefore a GA could be used to select between choices for any or all of the following:

- model inputs,
- number of hidden layers in the MLP,
- number of nodes in each hidden layer,
- nature of transfer functions at each node,
- connection structure between each node, and
- weights between each node.

The general evolutionary process is the same for all cases. A chromosome (string) is defined to represent the features of the MLP which are to be evolved. This chromosome may have a binary, real-valued or mixed form. Consider a case where the GA is evolving a feature of a MLP other than its connection weights. One possible algorithm (from many) is (Fig. 3.20):

   i. Decode the string into a MLP structure.
  ii. Initialise the weights.
 iii. Train the MLP using the backpropagation algorithm.
  iv. Determine the predictive accuracy (fitness) of the resulting MLP.
   v. Perform fitness-based selection to create a new population of genotypes.
  vi. Create diversity in the chromosomes of the new population (crossover/mutation operators).
 vii. Replace the old population with the new one.
viii. Repeat above steps until stopping criteria are met.

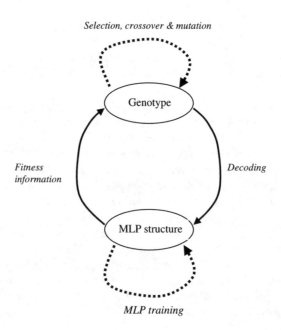

Fig. 3.20. An example of a GA-MLP hybrid. The hybrid uses evolutionary learning to uncover a good structure, but the weights are obtained using the backpropagation algorithm

Evolving the Selection of Inputs

If the intention is solely to evolve good sets of inputs the choice of representation is straightforward, a binary string N items long, where each element of the string is an indicator $(0,1)$ as to whether each of the N inputs is included or excluded. For each selection of inputs, the MLP is trained and an error measure obtained, which is then used to guide the evolutionary process in its search for the best set of inputs. A particular advantage of having the GA determine the selection of input variables is that it is possible to have a variable included through the action of the mutation operator, at all stages of the training process, even if it has been previously deselected. This may allow the testing of a new grouping of variables, not all of which were present the last time the (reselected) variable was included. This property could be very useful if there are subtle, non-linear interactions between the input variables.

Evolving the Weights

If the objective is to evolve the weights for a MLP of fixed structure and fixed inputs, the weights can be *directly encoded* onto a chromosome. In direct encoding, the network architecture is effectively encoded in a chromosome and it is possible to recreate the exact MLP from this (Fig. 3.21). At the commencement of the training process, the chromosome is decoded to give the weights for each connection in the resulting network. The network is then trained, and an error measure is obtained for it. Low errors represent high fitness, and the GA acts to select good sets of weights, and creates variants on these in an effort to uncover better sets of weights in future generations of the algorithm. Although the above method can be used to evolve weights for a MLP, it will not necessarily outperform more-specialist weight-adjustment techniques for training MLPs (such as quickprop [72]).[1] Weight evolution does have clear advantage in cases where backpropagation methods cannot be used, for example when a specialist error measure is required which is not a differentiable function, or in cases which backpropagation finds hard, for example multi-modal error surfaces.

One important consideration when developing hybrid GA-MLPs is that the choice of representation has implications for the optimal design of the diversity-generating operators in the evolutionary component of the hybrid. In particular, the crossover operator must try to avoid disrupting good elements (building blocks) of a MLP's design once they are uncovered. Learning in MLPs is *distributed* across weights. Hence insensitively applied crossover, where weight values from different MLPs are stochastically recombined, can produce poor results. If we take one specific case as an example, where the goal is to apply crossover to two MLPs with the same structure and inputs in an effort to evolve good weights for the MLP, one simple approach is to define a basic building block as being a hidden or output node along with all its incoming connection weights. Hence a child MLP structure can be obtained from two parent MLPs by applying a crossover operation, which swaps hidden or output nodes along with their incoming connection weights between

---

[1]The basic idea of quickprop is to use the second derivative of the error function to attempt a direct step to the error minimum. It assumes that the error surface is locally quadratic.

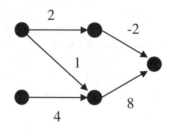

| 10100 | 11000 | 00000 | 10010 | 00100 | 10001 |

Fig. 3.21. Binary encoding of weights in a MLP. The first bit indicates whether the weight is negative (0) or positive (1), with the remaining four bits (reading from left to right) encoding the weight value. The weight 00000 indicates no connection between two nodes

the two parents. The mutation process can be applied to the same building block, whereby one or more of the hidden/output nodes of the child MLP are randomly selected, and a real-valued mutation is applied to the incoming connection weights of that node.

Although the above definition of a building block is intuitive, the efficiency of the crossover operator is impacted by the *permutation problem*, also known as the *competing conventions problem* [105, 155]. Consider two high-fitness MLPs which are identical, except that their input or hidden nodes are permutated in a different order. If these MLPs are encoded as a linear binary string, a crossover operation between the two MLPs is likely to be disruptive, producing child MLPs of lower fitness than either parent.

Evolving the Connection Structure

A more complex task is to use the GA to evolve the structure for a MLP. In these cases, the modeller is faced with selecting a way to represent the range of possible network structures in the chromosome. One way of encoding the connection topology is by means of a connection matrix. This matrix is of size $N * N$, where N is the maximum number of possible nodes in the network (Figs. 3.22 and 3.23). The connection matrix consists of (0,1) values, each indicating whether that connection is used or *turned on* in that network. This representation implicitly allows the GA to select the number of hidden nodes it wishes to use, as a hidden node which is not connected to any input or output nodes is effectively deselected. If N is large, the feasibility of employing direct encoding declines, as a connection matrix for N nodes will scale at a rate of $O(N^2)$. Another problem that can arise when MLP architectures are being evolved without any weight information is that of *noisy fitness* evaluation. This occurs because the fitness of the resulting phenotype also depends on the initialisation of the weights at the start of the (non-evolutionary) training process. It is possible that a good architecture could obtain a low fitness, just because of a poor weight initialisation, and therefore not be selected for subsequent generations. One way around this problem is to train the MLP using several different weight initialisations, and use the average of these results as the fitness of the genotype. Another way around this problem is to evolve both the architecture and the weights simultaneously (see Chapt. 12 for an example of this). Yao (1999) [224] provides a good review of the technical issues involved in creating hybrid GA-MLP systems.

Recent approaches to evolving neural networks include enforced sub-populations (ESP) [94, 95] and NEAT (neuroevolution of augmenting topologies) [203]. In ESP, rather than evolving chromosomes which encode complete networks, a series of separate sub-populations are evolved, each of which encodes the incoming and outgoing weights of a single (hidden) node. A complete network can be formed by selecting a node from each population, and an evolutionary process is applied to improve the quality of each sub-population over time. In the NEAT approach, both the topology and weights of the network are evolved starting from a minimal network size.

Indirect Encodings

Apart from direct encoding schemes, where some aspect of the MLP's structure is directly encoded onto a genotype, it is possible to generate MLPs using an indirect encoding. These encodings may specify a set of construction rules that are iteratively applied to produce the MLP. Examples of this approach include the *grammatical/cellular encoding* methods [99, 127]. In these methods, the basic building blocks of the network are encoded in the form of a grammar (a set of rules which can be applied to produce a structure, in this case a complete MLP), and MLPs are developed by stringing together the building blocks defined in the grammar. A detailed discussion of grammars, and how they can be used to develop structures as diverse as computer programs or trading systems, is provided in Chapt. 4 which discusses grammatical evolution.

In summary, the combination of an evolutionary algorithm with a MLP incorporates the complementary strengths of two distinct methodologies. The strengths of MLPs (parallel computation, universal approximator properties, robustness to noise, and adaptive learning) are combined with the global search potential of a GA. It is also possible to use a GA to integrate the prediction of a series of MLPs, perhaps each trained using different input

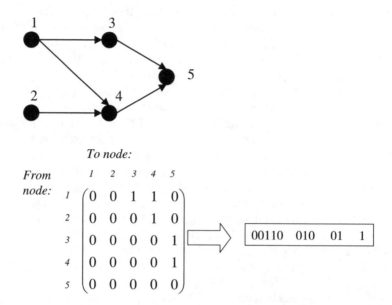

Fig. 3.22. A feedforward MLP connection matrix. The matrix can be concatenated into a binary string. Due to the feedforward architecture, the binary string only needs to contain the upper-right triangle of the matrix

data. The outputs from the MLPs could be integrated by applying different weights to the predictions of each individual MLP, using the GA to evolve the weighting factors. More generally, the combination could be of multiple systems which use a variety of predictive methodologies. Finally, although the above section discussed the potential for combining GAs and MLPs, many of the same concepts could be applied to evolve other forms of neural networks.

3.5 Applying EAs to Evolve Trading Rules

A key advantage of adopting an evolutionary approach to designing a trading system, is that it is capable of simultaneously evolving *both* the trading rule and good parameters for that rule. This provides a powerful extension to methods such as MLPs, where the modeller must implicitly select the rule structure through selection of which inputs will be used, and by selection of what model structure the MLP will have (how many layers, how many hidden layer nodes, etc.).

Let us examine the case where an EA is used to evolve good parameters for a prespecified trading rule. Suppose a trader wanted to develop a rule

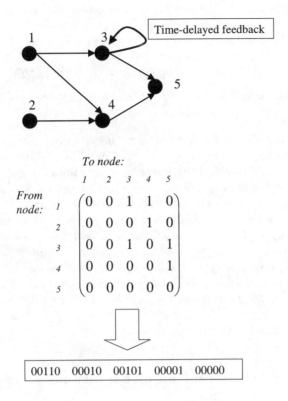

Fig. 3.23. A recurrent architecture encoded as a binary string. As feedback arcs are permitted, the binary string must encompass the entire connection matrix

based on the crossover of the moving average of two timeseries of historic price information for a financial asset. A simple rule could take the form

$$IF \left(\frac{x \text{ day moving average of price}}{y \text{ day moving average of price}} > z \right) \quad THEN \; buy \; ELSE \; sell$$

(no claim is made that a such a simple rule will generate profits!). The object is to determine the values of x, y and z which will produce the best results. A random population of binary strings can be initialised to encode real values for x, y and z. The fitness of each string can then be determined by translating the binary string into three real numbers, and using historical market information to determine what the fitness of the trading rule would have been in a historic period. Applying the usual evolutionary process already outlined, good values for the three parameters can be determined.

More generally, the EA can be used to evolve the trading rule as well as parameters for the rule. For example, suppose a modeller specifies a broad format for the desired rules a priori. An example could be:

$$IF\ [INDICATOR_1(time)\ (<,>,=)\ VALUE_1]$$

$$(AND,\ OR,\ NOT)\ [INDICATOR_2(time)\ (<,>,=)\ VALUE_2]$$

$$THEN\ (BUY,\ SELL,\ DO\ NOTHING)$$

In this rule format the EA is allowed to choose from multiple indicators (for example moving averages of price or trading volume data), associated time variables (or lags) for each indicator, trigger values for each indicator, the form of the mathematical operator between the indicator and the trigger value $(<,>,=)$, how the tests for each indicator should be linked (choice of AND, OR, NOT functions) and, finally, the trading signal to be generated for that rule. The choice of indicator, mathematical function, boolean function and trading signal can all be encoded as integers or as binary values which are decoded into integers. During the EA runs, different vectors of these values can be converted into trading rules, and the fitness of each rule can be determined by applying it to the training dataset. In turn, the differential fitnesses of the resulting trading rules drives the evolutionary process, leading to the uncovering of better trading rules over time.

3.6 Recent Developments in Evolutionary Computation

An important new direction in genetic algorithm research stems from the recognition of the limited scalability of the canonical GA when it is applied to problems of increasing difficulty. It has been recognised that the success of a GA is dependent upon facilitating the proper growth and mixing of building blocks, which is not achieved by problem-independent recombination operators [92, 214]. The algorithms emerging from this area of research are dubbed *competent GAs*. Competent GAs seek to perform a more intelligent search by respecting the functionally important linkages between the constituent components of a solution in order to prevent the disruption of potentially useful building blocks. As an example of this, consider a binary string encoding which is N bits long, where the first and the last bit must both be '1' if the string is to have high fitness. If basic single-point crossover is applied to two parents, one of which already has the correct ('1') value in these locations, it is quite possible that neither child will inherit the good genes from that parent. In other words, because no attention has been paid to the linkage structure between the elements of the string, the crossover operator has acted in a destructive manner. Ideally, the aim is to codesign the encoding representation and the diversity-generating operators in order to minimise this problem. More recently, the GP community is beginning to apply ideas from competent GA in designing GP algorithms [190, 196].

Earlier in the chapter we introduced a popular form of tree-based GP [131]. However, prior to its introduction a number of alternative representations had been adopted (for example see [45, 83, 84]) and since then a large variety of representations have been examined including graphs, linear structures, grammars and even hybridisations of these. Notable examples include linear GP [14, 162], PADO (Parallel Algorithm Discovery and Orchestration) [212], graph and linear-graph GP [120, 121], Cartesian GP [152], and grammar-based GP systems (e.g., [99, 106, 174, 218, 222]). While the issue of representation is not unique to GP, indeed more broadly it transcends machine learning as a whole, the question as to what makes a *good* representation for EC is an open one and some attempts are now being initiated to formalise this research [185]. In recent years there has been a great deal of research on schema theories for genetic programming, and it is being recognised that these theories demonstrate a commonality between the various representations adopted in EC, with GP schema theories being considered supersets of GA schema theory [138].

One hybrid representation combining linear chromosomes and grammars, grammatical evolution, has attracted notable interest in recent years [81, 174, 176, 177, 178, 220]. Grammatical evolution provides an elegant mechanism to ensure the all-important GP property of closure whilst allowing the removal of the restriction of programs to a single data type. In addition, grammatical evolution provides a convenient manner in which to incorporate domain knowledge into the representation of a solution through its inclusion in the input grammar.

3.7 Summary

This chapter presented an introduction to a family of algorithms inspired by an evolutionary metaphor, evolutionary algorithms. Specific EA instances were examined, namely genetic algorithms, differential evolution and genetic programming. Following two examples of how EAs can be applied for financial prediction purposes, we outlined some of the more recent developments in EC. Our presentation of evolutionary computation continues in Chap. 4 with an introduction to the grammatical evolution framework.

4

Grammatical Evolution

This chapter provides an introduction to grammatical evolution (GE), a form of grammar-based genetic programming which has been applied to a number of problem domains including the evolution of computer programs and financial prediction.

In addition to the standard set of evolutionary principles adopted in evolutionary computation, as described in Chap. 4, GE further extends the biological analogy by employing principles from genetics that have been uncovered by molecular biologists. The most significant of these is the adoption of a distinction between the genotype and phenotype similar to that which exists in Nature. That is, through a mapping process the genetic material (the genotype) contains the instructions that are used to control the development and day-to-day operation of a living organism (the phenotype). The molecules making up the genetic material (DNA) are distinct from the molecules responsible for the phenotype (proteins).

It is in this notion of a genotype-phenotype mapping that the use of a grammar is exploited. The grammar contains the rules governing how the development of the phenotype is conducted, and as such can contain domain knowledge biasing the form a phenotypic solution can take.

In this chapter we will introduce the underlying biological principles upon which GE is based, we provide an example of how GE operates, and draw the reader's attention to some of the more recent developments in GE.

4.1 Grammatical Evolution

GE is an evolutionary algorithm that can evolve *computer programs, rulesets* or more generally *sentences* in any language [166, 173, 174, 189]. Rulesets could be as diverse as a regression model or a trading system for a financial market. Rather than representing the programs as syntax trees, as in GP [131], a *linear genome* representation is used in conjunction with a grammar. Each individual, a variable-length binary string, contains in its codons (groups of 8 bits) the information to select production rules from a grammar.

4.1.1 Biological Analogy

The GE system is inspired by the biological process of generating a protein from the genetic material of an organism. Proteins are fundamental in the proper development and operation of living organisms and are responsible for traits such as eye colour and height [141].

The genetic material (usually DNA, deoxyribonucleic acid) contains the information required to produce specific proteins at different points along the molecule. For simplicity, consider DNA to be a string of building blocks called nucleotides, of which there are four, named A, T, G and C, for adenine, tyrosine, guanine and cytosine respectively. Groups of three nucleotides, called codons, are used to specify the building blocks of proteins. These protein building blocks are known as amino acids, and the sequence of these amino acids in a protein is determined by the sequence of codons on the DNA strand. The sequence of amino acids is very important as it plays a large part in determining the final three-dimensional structure of the protein, which in turn has a role to play in determining its functional properties.

In order to generate a protein from the sequence of nucleotides in the DNA, the nucleotide sequence is first transcribed into a slightly different format, a sequence of elements on a molecule known as RNA (ribonucleic acid). Codons within the RNA molecule are then translated to determine the sequence of amino acids that are contained within the protein molecule. The application of production rules to the non-terminals of the incomplete code being mapped in GE is analogous to the role amino acids play when being combined together to transform the growing protein molecule into its final functional three-dimensional form.

The result of the expression of the genetic material as proteins in conjunction with environmental factors is the phenotype. In GE, the phenotype is a sentence(s) in some language (e.g., a program in the C programming language) that is generated from the genetic material (the genotype). This is unlike the standard method of generating a solution (e.g., a program in the case of GE) directly from an individual in an evolutionary algorithm by explicitly encoding the solution within the genetic material. Instead, a many-to-one mapping process is employed.

Figure 4.1 compares the mapping processes employed in GE and biological organisms. Through the adoption of a genotype-phenotype mapping process coupled to the use of a grammatical representation, GE can take advantage of its modular framework in a number of ways. Benefits include:

- A separation of the search (binary strings) and solution spaces (sentences) which removes the necessity to exclusively adopt a simple variable-length

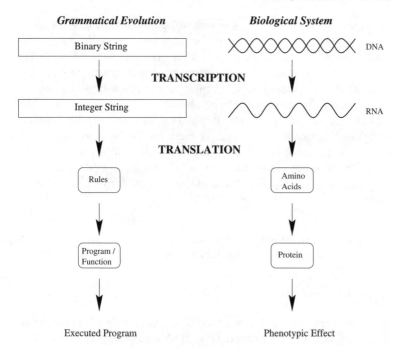

Fig. 4.1. A comparison between the grammatical evolution system and a biological genetic system. The binary string of GE is analogous to the double helix of DNA, each guiding the formation of the phenotype. In the case of GE, this occurs via the application of production rules to generate the terminals of the compilable program. In the biological case by directing the formation of the phenotypic protein by determining the order and type of protein subcomponents (amino acids) that are joined together

genetic algorithm (or even any evolutionary algorithm!) as the search engine. The search operators of the evolutionary algorithm themselves (e.g., the genetic operators of crossover and mutation) operate on an abstraction of the phenotype and as such do not have to take into consideration issues such as syntactic correctness of the phenotype, as the mapping process can be used to ensure this occurs automatically.

• An abstraction of a program's representation (a grammar) which can be used in a plugable manner to generate sentences in arbitrary languages.

- Efficiency gains for an evolutionary search are possible through the adoption of the many-to-one mapping and the degenerate genetic code. This can be achieved by allowing neutral evolution to occur. Neutral evolution occurs when there are changes at the genotype level which are neutral or nearly neutral with respect to the fitness of the phenotype. For example, by allowing the functionality of the phenotype to be preserved while changes to the genotype occur, the population can potentially traverse otherwise infeasible regions of the search space. A further consequence of the many-to-one mapping is the maintenance of genetic diversity within a population by allowing many different genotypes to represent the same phenotype, thus helping to prevent loss of genetic material and premature convergence.

We will now present an overview of how the genotype-phenotype mapping process occurs in GE through the mapping of a sample individual.

4.1.2 Mapping Process

When tackling a problem with GE, a suitable BNF (Backus-Naur form) grammar definition must initially be defined. The BNF can be either the specification of an entire language or, perhaps more usefully, a subset of a language geared towards the problem at hand.

In GE, a BNF definition is used to describe the output language to be produced by the system. BNF is a notation for expressing the grammar of a language in the form of production rules. BNF grammars consist of *terminals*, which are items that can appear in the language, e.g., binary boolean operators **and**, **or**, **xor** and **nand**, unary boolean operators **not**, constants, **true** and **false**, etc., and *non-terminals*, which can be expanded into one or more terminals and non-terminals.

For example the grammar below can be used to generate boolean expressions, and `<expr>` can be transformed into one of three rules. It can become either (`<expr> <biop> <expr>`), `<uop> <expr>`, or `<bool>`. A grammar can be represented by the tuple $\{N, T, P, S\}$, where N is the set of non-terminals, T the set of terminals, P a set of production rules that maps the elements of N to T, and S is a start symbol which is a member of N. When there are a number of productions that can be applied to one element of N the choice is delimited with the '|' symbol. For example

```
N = {<expr>, <biop>, <uop>, <bool>}
T = {and, or, xor, nand, not, true, false, (, )}
S = {<expr>}
```

And P can be represented as:

```
(A) <expr> ::= (<expr> <biop> <expr>)      (0)
           | <uop> <expr>                  (1)
           | <bool>                        (2)

(B) <biop> ::= and        (0)
           | or           (1)
           | xor          (2)
           | nand         (3)

(C) <uop> ::= not

(D) <bool> ::= true       (0)
           | false        (1)
```

The code produced will consist of elements of the terminal set T. The grammar is used in a developmental approach whereby the evolutionary process evolves the production rules to be applied at each stage of a mapping process, starting from the start symbol, until a complete program is formed. A complete program is one that is comprised solely from elements of T.

As the BNF definition is a plug-in component of the system, it means that GE can produce code in any language thereby giving the system a unique flexibility. For the above BNF grammar, Table 4.1 summarises the production rules and the number of choices associated with each.

Table 4.1. The number of choices available from each production rule

Rule Number	Choices
A	3
B	4
C	1
D	2

The genotype is used to map the start symbol onto terminals by reading codons of 8 bits to generate a corresponding integer value, from which an appropriate production rule is selected by using the following mapping function:

$$Rule = c \mod r$$

where c is the codon integer value, and r is the number of rule choices for the current non-terminal symbol.

Consider the following rule from the above sample grammar, i.e., given the non-terminal `<biop>`, which describes the set of binary operators that can be used, there are four production rules to select from.

```
(B) <biop> ::= and        (0)
           | or           (1)
           | xor          (2)
           | nand         (3)
```

If we assume the codon being read produces the integer 6, then

$$6 \; mod \; 4 = 2$$

would select rule (2) `xor`. That is, `<biop>` is replaced with `xor`. Each time a production rule has to be selected to transform a non-terminal, another codon is read. In this way the system traverses the genome.

During the genotype-to-phenotype mapping process, it is possible for individuals to run out of codons, and in this case the *wrap* operator is applied which results in returning the codon reading head back to the first codon in the individual. As such codons are reused when wrapping occurs. This is quite an unusual approach in evolutionary algorithms as it is entirely possible for certain codons to be used two or more times. This technique of wrapping the individual draws inspiration from the gene-overlapping phenomenon that has been observed in many organisms [141].

In GE each time the same codon is expressed it will always generate the same integer value, but, depending on the current non-terminal to which it is being applied, it may result in the selection of a different production rule. This feature is referred to as *intrinsic polymorphism*. What is crucial, however, is that each time a particular individual is mapped from its genotype to its phenotype the same output is generated. This is the case because the same choices are made each time. It is possible that an incomplete mapping could occur, even after several wrapping events, and typically in this case the mapping process is aborted and the individual in question is given the lowest possible fitness value. The selection and replacement mechanisms then operate accordingly to increase the likelihood that this individual is removed from the population.

An incomplete mapping could arise if the integer values expressed by the genotype were applying the same production rules repeatedly. For example, consider an individual with three codons, all of which specify rule (0) from:

```
(A) <expr> ::= (<expr> <biop> <expr>)    (0)
           | <uop> <expr>                (1)
           | <bool>                      (2)
```

Even after wrapping, the mapping process would be incomplete and would carry on indefinitely unless terminated. This occurs because the nonterminal

`<expr>` is being mapped recursively by production rule (0), i.e., it becomes (`<expr> <biop> <expr>`). Therefore, the leftmost `<expr>` after each application of a production would itself be mapped to a (`<expr> <biop> <expr>`), resulting in an expression continually growing as follows: ((`<expr> <biop> <expr>`) `<biop> <expr>`) followed by (((`<expr> <biop> <expr>`) `<biop> <expr>`) `<biop> <expr>`) and so on.

Such an individual is dubbed invalid as it will never undergo a complete mapping to a set of terminals. For this reason an upper limit on the number of wrapping events that can occur is imposed. During the mapping process, therefore, beginning from the left-hand side of the genome codon integer values are generated and used to select rules from the BNF grammar, until one of the following situations arise:

i. A complete program is generated. This occurs when all the non-terminals in the expression being mapped are transformed into elements from the terminal set of the BNF grammar.

ii. The end of the genome is reached, in which case the *wrapping* operator is invoked. This results in the return of the genome reading frame to the left-hand side of the genome once again. The reading of codons will then continue unless an upper threshold representing the maximum number of wrapping events is reached during this individual's mapping process.

iii. In the event that a threshold on the number of wrapping events is reached and the individual is still incompletely mapped, the mapping process is halted, and the individual is assigned the lowest-possible fitness value.

To reduce the number of invalid individuals being passed from generation to generation, a steady-state replacement mechanism is commonly employed. One consequence of the use of a steady-state method is its tendency to maintain fit individuals at the expense of less fit, and, in particular, invalid individuals. Alternatively, a repair strategy can be adopted, which ensures that every individual results in a valid program. For example, in the case that there are non-terminals remaining after using all the genetic material of an individual (with or without the use of wrapping) default rules for each non-terminal can be prespecified that are used to complete the mapping in a deterministic fashion. Another strategy is to remove the recursive production rules that cause an individual's phenotype to grow, and then to reuse the genotype to select from the remaining non-recursive rules.

4.1.3 Mapping Example

Consider the following genome, represented as a series of integer-valued codons:

42 22 6 104 70 31 13 4 25 9 3 86 44 48 3 27 4 111 56 2

We will demonstrate the application of this genome to the grammar presented below, which could be used to generate a simplified trading system. The trading system has ten possible input variables (var0 → var9), and three trading signals can be generated by the system: buy, sell, or do nothing. The start symbol (<S>) for the grammar from which the mapping process commences is the non-terminal <tradingrule>.

```
<S> ::= <tradingrule>

<tradingrule>  ::= if(<signal>) {<trade>;} else {<trade>;}

<signal> ::= <value> <relop> <var>
           | (<signal>) AND (<signal>)
           | (<signal>) OR (<signal>)

<value> ::= <int-const> | <real-const>

<relop> ::= <= | >=

<trade> ::= buy
          | sell
          | do-nothing

<int-const> ::= <int-const><int-const>
              | 1 | 2 | 3 | 4 | 5 | 6 | 7 | 8 | 9

<real-const> ::= 0.<int-const>

<var> ::= var0 | var1 | var2 | var3 | var4
        | var5 | var6 | var7 | var8 | var9
```

As there is only one production rule for <tradingrule> it is automatically replaced with its right-hand side (42 *mod* 1 = 0). Our developing trading rule becomes:

```
if(<signal>) {<trade>;} else {<trade>;}
```

Taking the left-most non-terminal <signal> there are three possible replacements. The codon reading head moves one codon to the right and now is above '22'.

42 **22** 6 104 70 31 13 4 25 9 3 86 44 48 3 27 4 111 56 2

Reading the next codon value determines that we use the second production rule (22 *mod* 3 = 1), which allows the logical AND of two or more infix expressions . This results in the following:

```
if((<signal>) AND (<signal>)) {<trade>;} else {<trade>;}
```

Again, taking the left-most non-terminal <signal> there are three choices. The next codon value dictates that we replace this <signal> with a single relational expression (i.e., $6 \bmod 3 = 0$) giving:

```
if(((<value> <relop> <var>) AND (<signal>)) {<trade>;}
else {<trade>;}
```

According to the next codon value ($104 \bmod 2 = 0$) the non-terminal <value> is replaced with an integer (<int-const>). The development of <int-const> proceeds as follows:

$70 \bmod 10 = 0$ results in the constant becoming a double-digit integer (i.e., <int-const><int-const>). The left-most <int-const> becomes 1 by $31 \bmod 10 = 1$ and the next <int-const> becomes 3 by $13 \bmod 10 = 3$.

The developing trading rule now has the form:

```
if((13 <relop> <var>) AND (<signal>)) {<trade>;}
else {<trade>;}
```

<relop> is replaced with <= according to $4 \bmod 2 = 0$, and the following <var> becomes var5 (i.e., $25 \bmod 10 = 5$) resulting in:

```
if((13 <= var5) AND (<signal>)) {<trade>;}
else {<trade>;}
```

By $9 \bmod 3 = 0$ the next <signal> becomes a single relational expression (<value><relop><var>) and its <value> component is replaced with a <real-const> according to $3 \bmod 2 = 1$. The <real-const> is expanded to become 0.64 by reading the following two codon values ($86 \bmod 10 = 6$ and $44 \bmod 10 = 4$). The relational operator becomes <= ($48 \bmod 2 = 0$), and var3 replaces <var> ($3 \bmod 10 = 3$).

Finally, we use the remaining codons to determine the fate of the two <trade> non-terminals in:

```
if((13 <= var5) AND (0.64 <= var3)) {<trade>;}
else {<trade>;}
```

The left-most <trade> becomes a buy() according to $27 \bmod 3 = 0$, while the final <trade> is replaced with sell() (i.e., $4 \bmod 3 = 1$). The position of the codon reading head is illustrated by the bold character below.

42 22 6 104 70 31 13 4 25 9 3 86 44 48 3 27 4 111 56 2

The three leftover codons are unused during the mapping process and are simply ignored, and consequently are referred to as introns as they do not impact on the function of the phenotype. The final trading system is then comprised of the following rule:

```
if((13 <= var5) AND (0.64 <= var3)) {buy();}
else {sell();}
```

The variables (var0 to var9) could represent a selection of elements of information drawn from fundamental analysis of an industry sector; for example, var5 could be a price/earnings ratio for a company, and var3 could represent a company's sales growth over the past 3 years. Hence, the above rule buys companies which have a P/E ratio of more than 13, and where the level of sales growth over the past three years exceeds 64%. Of course, successful real-world filter rules for trading would not be as simple as this, and would typically contain multiple conditions.

4.2 Mutation and Crossover in GE

As the search process in the canonical GE is driven by a variable-length GA (Fig. 4.2), the standard genetic operators such as crossover, mutation and duplication can be applied to the underlying genotypic representation irrespective of the form of the phenotype. That is, an unconstrained evolutionary search can be conducted at the genotypic level, unlike in GP where the search operators have to be designed to ensure that syntactically correct programs are created. GE's genotype-phenotype mapping process automatically ensures that syntactically legal programs are generated. As a result of this genotype-phenotype mapping, it is interesting to examine the effects the operators of mutation and crossover have on the phenotypic program.

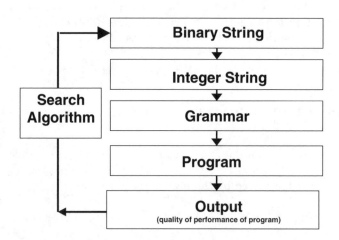

Fig. 4.2. Grammatical evolution embeds a populational search process in which the quality of the phenotype drives the search process. The diversity-generation process acts on the underlying genotype

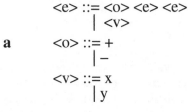

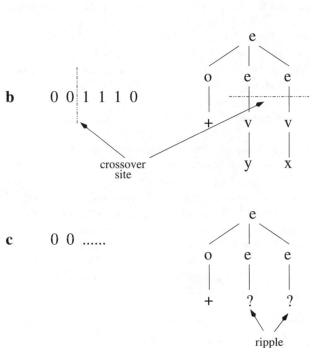

Fig. 4.3. An illustration of ripple crossover in grammatical evolution using the chromosome (represented as rule choices) (b) and its corresponding derivation tree, which is produced as a result of the grammar (a). The site of one-point crossover is indicated (b) on the chromosome and the derivation tree. The resulting derivation tree ripple sites are indicated with '?' (c)

In the case of mutation events at the codon level, the integer value of a codon will be modified. However, depending on the context (the non-terminal) to which the codon is applied the change in its value may or may not result in a change in the choice of production rule applied. For example, given the non-terminal <value> with two possible rules (see below), a binary codon value of 00001010 (decimal value 10) that results in <value> being replaced with <int-const> after bit mutation to 00001011 (decimal value 11) results in the use of <real-const>:

```
<value> ::= <int-const>
          | <real-const>
```

On the other hand, a mutation event that results in 00001011 becoming 01001011 (decimal value 75) leaves <value> being replaced with <real-const>. This type of mutation event is referred to as neutral mutation as it has no effect on the phenotype's functionality (fitness). In a similar fashion to the degenerate biological genetic code in which different codons can represent the same amino acid, different codons in GE can represent the same choice of production rule when used in the same non-terminal context.

A standard one-point crossover event on a GE chromosome results in the right-hand sides of the parental chromosomes undergoing a simple swap in a standard GA fashion. However, the impact on the phenotype may not be so simple. One-point crossover in GE has been called *ripple crossover* due to the effect that can arise on the resulting derivation sequence. Figure 4.3 illustrates this process. During the genotype-phenotype mapping in GE the resulting derivation sequence can be represented as a derivation tree. The impact of a crossover event on the genotype can be seen in the example derivation tree (Fig. 4.3). A number of *ripple sites* at different locations on the derivation tree are created once the genetic material on the right-hand side of the crossover site is removed, the result being that the codons swapped over from the second parent are used to complete the derivation sequence at these incomplete points. When compared to sub-tree crossover in GP (see Chap. 3) ripple crossover has a completely different behaviour. Instead of modifying a single sub-tree as in GP, multiple sub-trees can be changed as a consequence of applying ripple crossover in GE.

4.3 Recent Developments in GE

Grammatical evolution has received considerable attention since its introduction and there is a wide literature on the topic including [174, 176, 177, 178].[1] Some of the more recent developments are focused on the various components (Fig. 4.4) of the GE approach including the search engine, the grammar, and the mapping process itself. Some of the significant advances are detailed below.

4.3.1 Search Engine

Alternative search engines to the variable-length GA have been adopted. For example, particle swarm and differential evolution algorithms have been used to create grammatical swarm [168] and grammatical differential evolution algorithms, respectively. The grammatical swarm (GS) algorithm has shown particular promise and demonstrates a social learning approach to program

---

[1]See also http://www.grammatical-evolution.org

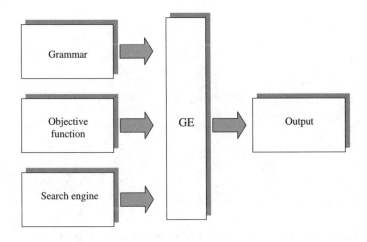

Fig. 4.4. Modular structure of grammatical evolution

generation dubbed *social programming*. A short description of the GS algorithm is provided in Chap. 5.5.

4.3.2 Meta-grammars

The adoption of a meta-grammar, a grammar that describes the construction of another grammar, has been investigated [175]. This grammatical evolution by grammatical evolution (GE^2) approach allows the grammar itself to be evolved and has been shown to be particularly effective in dynamic environments.

Exploiting this meta-grammar approach, a modular grammar-based GA, mGGA (the meta-grammar genetic algorithm), has been described [167]. The mGGA can automatically compose building blocks of symbols that can be reused in a modular fashion to construct a solution in a more efficient manner. By exploiting modularity within a solution, the length and therefore complexity of the genotype encoding the solution can be reduced, thus improving the scalability of the algorithm to tackle harder instances of problems. An example meta-grammar to generate 8-bit solutions with the mGGA is presented below:

```
<g> ::=
        "<bitstring> ::=" <reps>
            "<bbk4> ::=" <bbk4>
            "<bbk2> ::=" <bbk2>
            "<bbk1> ::=" <bbk1>
            "<bit> ::=" <val>

<bbk4> ::= <bbk4t>
        | <bbk4t> "|" <bbk4>
```

```
<bbk2>  ::= <bbk2t>
          | <bbk2t> "|" <bbk2>
<bbk1>  ::= <bbk1t>
          | <bbk1t> "|" <bbk1>
<bbk4t> ::= <bit><bit><bit><bit>
<bbk2t> ::= <bit><bit>
<bbk1t> ::= <bit>
<reps>  ::= <rept>
          | <rept>  "|" <reps>
<rept>  ::= "<bbk4><bbk4>"
          | "<bbk2><bbk2><bbk2><bbk2>"
          | "<bbk1><bbk1><bbk1><bbk1><bbk1><bbk1><bbk1><bbk1>"
<bit>   ::= "<bit>"
          | 1
          | 0
<val>   ::= <valt>
          | <valt> "|" <val>
<valt>  ::= 1
          | 0
```

Building blocks of size 1, 2, 4 and 8 bits are specified to be components of the solution grammar output as the result of mapping the above meta-grammar. For each building block size there can be many different building block instances represented as choices for that building block size in the solution grammar. An example bitstring solution grammar that could be produced by the above meta-grammar is provided below:

```
<bitstring> ::= <bit>11<bit>00<bit><bit>
              | <bbk2><bbk2><bbk2><bbk2>
              | 11011101
              | <bbk4><bbk4>
              | <bbk4><bbk4>

<bbk4>  ::= <bit>11<bit>
          | 000<bit>

<bbk2>  ::= 11
          | 00
          | <bit>1

<bbk1>  ::= 0
          | 0

<bit>   ::= 1 | 0 | 0 | 1
```

We can see in the above solution grammar that there are five possible building blocks of size eight (`<bitstring>`), two possible building block types of size four (`<bbk4>`) and three possible building blocks of size two (`<bbk2>`). Modularity exists above in the ability to specify the size and content (or partial content) of a building block through its incorporation into the solution grammar. This building block can then be repeatedly reused in the generation of the phenotype. An additional mechanism for reuse is through the wrapping operator of grammatical evolution. During the mapping process if we reach the end of the genotype and the construction of our phenotype is incomplete, we can invoke the wrapping operator to move our reading head back to the first codon in the genome. This allows the reuse of rule choices if the codons are used in the same context.

4.3.3 πGE

The GE mapping process can be divided into a number of sub-components including the transcription and translation processes as outlined in Fig. 4.1. The πGE variant of GE replaces the translation process to allow evolution to specify the order in which production rules are mapped as opposed to the strict depth-first left-to-right mapping of the standard GE algorithm [170]. Given the trading rule example grammar, if during the mapping process we have the following developing trading rule then in the standard GE mapping process the first non-terminal <value> will be expanded next.

```
if(<value> <relop> <value>) {
    <trade>;
}
else{
    <trade>;
}
```

However, in πGE we use the genotype to dictate firstly which non-terminal from the five present to expand next, before deciding which production rule to apply to this non-terminal. The genome of an individual in πGE is different in that there are two components to each codon. That is, each codon corresponds to the pair of values (*nont*, *rule*). To decide which non-terminal (NT) to expand the *nont* value of the next codon is read and used in the following mapping function

$$NT = c_{nont} \ \% \ n$$

where c_{nont} is the *nont* value of the current codon and n is the number of non-terminals present. To decide which production rule (R) is to be applied to the selected non-terminal the standard GE mapping function uses the *rule* component of the current codon as follows

$$R = c_{rule} \ \% \ r$$

where c_{rule} is the *rule* value of the current codon and r is the number of production rule choices for this non-terminal.

πGE has shown significant performance gains over the standard GE algorithm on a number of benchmark problem instances.

4.3.4 Applications and Alternative Grammars

The flexibility provided by the plugability of the grammar makes the grammatical evolution approach amenable to application across a broad range of problem domains. It also provides an easy way to encode domain knowledge. When the user is unsure of the exact form a solution might take, the use of a meta-grammar can remove this system parameter from the user, allowing the search process to learn what form the grammar should adopt.

In addition to meta-grammars various alternative grammars have been explored including the use of attribute grammars to introduce semantic and context-sensitive information [41, 172]. The ability to encode context-sensitive and semantic information into a grammar opens up a whole range of problems that cannot be tackled efficiently with a context-free language.

Grammars have also been used as an alternative approach to specify the creation of constants for genetic programming using digit concatenation and a more persistent alternative to ephemeral random constants [58]. More recently meta-grammars have also been adopted for constant creation [60].

4.4 Summary

This chapter presented an introduction to grammatical evolution, one of the methodologies inspired by an evolutionary metaphor. In Part III of this book we present a number of case studies that illustrate the use of grammatical evolution to generate rules for index trading, intra-day trading, and foreign exchange trading; for the prediction of corporate failure; and finally for the classification of bond ratings. We now continue our exposition of biologically inspired methodologies with examples of social learning algorithms based on swarm intelligence.

5

The Particle Swarm Model

One model of social learning which has attracted interest in computer science in recent years is drawn from a swarm metaphor. Two popular variants of swarm models exist, those inspired by the flocking behaviour of birds or the sociological behaviour of a group of people, and those inspired by the behaviour of social insects such as ant colonies (introduced in Chap. 6). The essence of these systems is that they exhibit flexibility, robustness and self-organisation [23]. Although the systems can exhibit remarkable coordination of activities between individuals, this coordination does not stem from a centre of control or a *directed* intelligence, rather it is self-organising and emergent. This chapter introduces the *particle swarm optimisation* (PSO) algorithm which has been inspired by social learning processes.

5.1 PSO Algorithm

The PSO algorithm was introduced by Kennedy and Eberhart [124] and is described in detail in [126]. In PSO, a swarm of particles, which encode solutions to the problem of interest, move (*fly*) in an n-dimensional search space in an attempt to uncover better solutions. Each of the particles has two associated properties, a current position and a velocity. Each particle has a memory of the best location in the search space that it has found so far (*pbest*), and knows the best location found to date by all the particles in the population (*gbest*). At each step of the algorithm, particles are displaced from their current position by applying a velocity (or gradient) vector to them. The magnitude and direction of their velocity at each step is influenced by their velocity in the previous iteration of the algorithm, simulating momentum, and the location of a particle relative to the location of its pbest and the gbest. Therefore, at each step, the size and direction of each particle's move is a function of its own history (experience), and the social influence of its peer group. Figure 5.1 provides a flowchart of the PSO algorithm. Several variants of the PSO algorithm exist. The algorithm for the canonical continuous version of PSO is as follows:

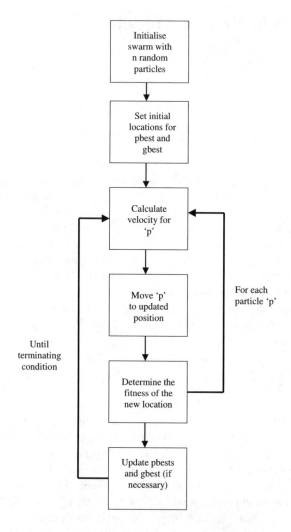

Fig. 5.1. A flowchart of the PSO algorithm

i. Initialise each particle in the population by randomly selecting values for its location and velocity vectors.

ii. Calculate the fitness value for each particle.

iii. Set pbest for each particle to its current location, and determine the location of gbest.
iv. For each particle, calculate its velocity using (5.1).
v. Update the position of each particle using (5.3).
vi. Calculate the fitness value of each particle. If the current fitness value for a particle is greater than its best previous fitness, then revise the location of *pbest*.
vii. After all the particles have been updated, determine the location of the particle with the highest fitness and revise *gbest* if necessary.
viii. Repeat steps iv-vii until stopping criteria are met.

Each particle i has an associated current position in search space x_i, a current velocity v_i, and a personal best position in search space y_i. During each iteration of the algorithm, the location and velocity of each particle are updated using (5.1-5.3). Assuming that a function f is to be maximised, that the swarm consists of n particles, and that r_1, r_2 are drawn from a uniform distribution in the range (0,1), the velocity update is as follows:

$$v_i(t+1) = Wv_i(t) + c_1r_1(y_i - x_i(t)) + c_2r_2(\hat{y} - x_i(t)) \qquad (5.1)$$

where $\hat{y}$ is the location of the global-best solution found by all the particles. A variant on the canonical particle swarm algorithm is to use a local best point (*lbest*) rather than a global best point (*gbest*). In the local version of the algorithm, each particle is considered to be linked to a subset of the population of particles, and this linkage structure is fixed at the beginning of the optimisation process. The term lbest replaces gbest in (5.1), and lbest represents the best location found by any particle in that local group.

At the start of the algorithm, the pbest for each particle is set at its initial location, and gbest is set to the location of the best of the pbests. In each iteration of the algorithm, a particle is stochastically accelerated towards its previous best position and towards a neighbourhood (global) best position, thereby forcing particles to continually search in the most-promising regions found so far in the solution space. The weight coefficients c_1 and c_2 control the relative impact of the pbest and gbest locations on the velocity of a particle. Low values for c_1 and c_2 allow each particle to explore far away from already uncovered good points, high values of the parameters encourage more intensive search of regions close to these points. The random coefficients r_1 and r_2 ensure that the algorithm is stochastic. A practical effect of r_1 and r_2, is that neither the individual nor the social learning terms are always dominant.

Each component of the velocity vector v_i is restricted to a range $[-v_{max}, v_{max}]$. The values chosen for $[-v_{max}, v_{max}]$ can have an important effect on the efficiency of the algorithm. Small values can result in insufficient exploration of the search space, while large values can result in particles moving past good solutions. The value of v_{max} is typically set in the range $k * x_{max}$, where $0 < k < 1$. Although this velocity restriction does not necessarily bound the particles into the range of maximum allowable values for each x_i during each iteration of the algorithm, the oscillation of the particles inside and outside the allowable range plays an important part in the search process of the swarm. Experience suggest that the particles should not be constrained to remain within x_{max} as the algorithm runs.

W represents a momentum coefficient which controls the impact of a particle's prior-period velocity on its current velocity. Higher values of the momentum term encourage the search of diverse regions. Typically, the value of W is decreased gradually during the search process, in an effort to encourage more-intensive local search of already discovered good regions, and to help the swarm converge. A simple method to achieve this is:

$$W = w_{max} - \frac{w_{max} - w_{min}}{max_{iter}} * curr_{iter} \qquad (5.2)$$

where w_{max} and w_{min} are the initial and final weight values, respectively (for example, 1.0 and 0.2), max_{iter} is the maximum number of iterations of the PSO algorithm, and $curr_{iter}$ is the current iteration number.

Once the velocity update for particle i is determined, its position is updated (5.3) and pbest ($y_i(t+1)$) is updated if necessary using (5.4-5.5).

$$x_i(t+1) = x_i(t) + v_i(t+1) \qquad (5.3)$$

$$y_i(t+1) = y_i(t) \text{ if } f(x_i(t+1)) \leq f(y_i(t)), \qquad (5.4)$$

$$y_i(t+1) = x_i(t+1) \text{ if } f(x_i(t+1)) > f(y_i(t)) \qquad (5.5)$$

After all particles have been updated a check is made to determine whether gbest needs to be updated:

$$\hat{y} \in (y_0, y_1, ..., y_n) | f(\hat{y}) = \max \left(f(y_0), f(y_1), ..., f(y_n) \right) \qquad (5.6)$$

Figure 5.2 provides visual intuition on the workings of the algorithm. A particle is located at position $x_i(t)$ at time t, and has a velocity of $v_i(t)$. The position of the particle at time $t+1$ is determined by $x_i(t) + v_i(t+1)$, and $v_i(t+1)$ is obtained by a stochastic blending of $v_i(t)$, an acceleration towards gbest (v_{gbest}) and an acceleration towards pbest (v_{pbest}).

5.1.1 Constriction Coefficient Version of PSO

An alternative method for controlling the magnitude of the velocity update step was proposed in [42], the *constriction coefficient*. In this method, the velocity update step is altered to the following:

$$v_i(t+1) = \chi(v_i(t) + c_1 r_1 (y_i - x_i(t)) + c_2 r_2 (\hat{y} - x_i(t))) \qquad (5.7)$$

where χ is the constriction coefficient (the momentum coefficient is dropped). The value of the constriction coefficient is calculated as $\chi = \frac{2}{|2 - c - \sqrt{(c^2 - 4c)}|}$,

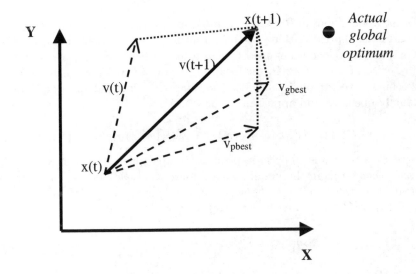

Fig. 5.2. Diagram of the particle position update process

where $c = c_1 + c_2$, and $c > 4$, the choice of these values being motivated to help ensure that the swarm converges to a small region of the search space. A common value for $\chi = 0.7298$, resulting from values of $c_1 = c_2 = 2.05$.

5.1.2 Parameter Settings for PSO

Relatively few parameters need to be tuned by the modeller in the canonical PSO algorithm. Common values for the key parameters include:

- Number of particles: typically 20-50.
- Range of values for particles: problem dependent.
- v_{max}: This determines the maximum change a particle can make in any direction during one iteration of the process. Typically it is set to the range of the allowable values for that parameter. For example, if the second element (parameter) of the solution is constrained to lie in [-5, 5] then v_{max} for that item is 10.
- Learning factors: c_1 and c_2. Generally, each is small in size and selected in the range [0,4], with $c_1 = c_2$. Popular values for each factor, for the non-constriction version of the algorithm, are in the range 1.5 to 2 [113].

5.2 Discrete PSO

In some applications of PSO, the required solutions have a binary rather than a continuous representation. The best-known version of binary PSO, *BinPSO* [125], converts the continuous PSO algorithm to one which operates on binary representations. In BinPSO, the location (x_i) of each particle i is represented as a binary vector of 0s and 1s. The adapted velocity-update equation is virtually unchanged in appearance from (5.1):

$$v_i^j(t+1) = (Wv_i^j(t) + c_1 r_1(y_i^j - x_i^j(t)) + c_2 r_2(\hat{y}^j - x_i^j(t))) \qquad (5.8)$$

where x_i^j is the value (0 or 1) in position j of particle i's location vector. All of the other terms in the update equation are as already defined in (5.1). To ensure that each element of the vector $\mathbf{v_i(t+1)}$ is binary, a sigmoid transformation is performed on each element j of $\mathbf{v_i(t+1)}$:

$$Sig(v_i^j(t+1)) = \frac{1}{1 + exp(-v_i^j(t+1))} \qquad (5.9)$$

Finally, these values are used to determine the values of each element of $x_i(t+1)$, by comparing each element of $Sig(\mathbf{v_i(t)})$ with a random number drawn from $U(0,1)$:

$$\text{If } U(0,1) < Sig(v_i^j(t+1)), \text{ then } x_i^j(t+1) = 1; \text{ else } x_i^j(t+1) = 0 \qquad (5.10)$$

Although the above velocity-update equation looks similar to that for the canonical continuous PSO, it has a quite a different interpretation in BinPSO. The vector v_i is interpreted as particle i's predisposition to set the value in each of the n binary elements of its position vector equal to 1. The higher the value of v_i^j for an individual element of i's position vector, the more likely that $x_i^j = 1$, with lower values of v_i^j favouring the choice of $x_i^j = 0$. $Sig(v_i^j)$ represents the probability of bit x_i^j taking the value 1 [125]. Therefore, if $Sig(v_i^j) = 0.3$ there is a 30% chance that $x_i^j = 1$, and a 70% chance it is zero.

5.3 Comparing PSO and the GA

The PSO algorithm bears similarity to other biologically inspired optimising algorithms. Like the GA, it is population-based, it is typically initialised with a population (swarm) of random encodings of solutions, and search proceeds by updating these encodings over a series of generations (iterations). Unlike the GA, PSO has no explicit selection process as all particles persist over time. Instead a *memory* in the form of gbest/lbest is substituted for selection.

It is possible to adapt the canonical PSO algorithm to incorporate an explicit selection mechanism. A simple replacement strategy would be to drop the poorest $x\%$ of particles after each iteration of the algorithm, replacing them with newly created random particles. A more sophisticated method is to periodically drop low-fitness particles, replacing their location and velocity vectors with those of higher-fitness particles in the current population, while leaving the pbest information for the replaced particle unchanged [10]. This has the affect of intensifying the search in a current good region, while maintaining a memory of historic high-fitness locations.

The communication or information-sharing mechanism of PSO differs from that of the GA. In the GA, the communication is between two solutions (the parents). In PSO the communication is between the current solution, its pbest and the gbest/lbest. Therefore, the communication process embeds both memory and peer-learning.

5.4 MLP-Swarm Hybrids

As already seen, search algorithms such as the GA can be used to uncover good weight vectors, choices of inputs, and network structures for MLP and other NN models. In a similar fashion, the particle swarm algorithm can be used to search for good choices for these items.

Suppose the intention was to uncover a good set of weights for an MLP of a fixed 6-3-1 (input-hidden-output node) structure. This MLP has 21 (6*3+3*1) connections, and the weights for each of these can be represented as a 21-dimension real-valued vector. In using PSO to uncover good weight vectors, a series of initial weight vectors (particles) could be generated randomly. Each weight vector can be decoded into an MLP structure, and the error for the resulting MLP determined by passing the training dataset through the network. The resulting model-fit error is a measure of the particle's fitness (low errors representing a higher fitness). Evidence suggests that the results obtained using this method are competitive with those obtained using the backpropagation algorithm [100]. The use of particle swarm to construct MLP models can be extended to include the search for good inputs, for good network structures, and for good slope values for the transfer functions [68].

5.5 Grammatical Swarm

In addition to the hybridisation of particle swarm algorithms with MLPs, they have also been combined with a grammatical evolution genotype-phenotype mapping [168, 169]. The resulting grammatical swarm (GS) algorithm uses PSO to search for computer programs, or more generally sentences in a user-specified language.

In GS each particle or real-valued vector represents choices of program construction rules specified as production rules of a Backus-Naur form grammar. The particle update equations are as described earlier for the continuous particle swarm algorithm with additional constraints placed on the velocity and dimension values. Velocities are bound to the range $\pm V_{max} = 255$, and each dimension is bound to the range 0 to 255. The real-valued dimension values are rounded up or down to the nearest integer, and the standard GE mapping function ($R = c$ Mod r, where R is the selected production rule, c is the codon integer value, and r is the number of production rules to choose from) is used.

Unlike its GE or GP counterparts, which predominantly use crossover-driven search coupled with selection, GS does not use explicit crossover or selection to generate programs. Instead the search process is driven by the movement of particles which are influenced by personal and social knowledge in the form of the positions of the *gbest* (or *lbest*) particle and the particle's own *pbest* position. The performance of GS has been compared to GE across a number of benchmark problems with encouraging results, suggesting that program generation using a social programming approach such as GS is a possible alternative to more traditional (GA-driven) genetic programming algorithms.

5.6 Example of a Financial Application of PSO

In Chap. 3 it was seen how a GA could be used to evolve rules for a trading system where the general format of these rules is specified by the modeller. An example of a possible rule is:

IF [INDICATOR$_1$(time) (<,>,=) VALUE$_1$]

(AND, OR, NOT) [INDICATOR$_2$(time) (<,>,=) VALUE$_2$]

THEN (BUY, SELL, DO NOTHING)

The creation of a good trading rule represents a search problem. Of all the possible choices of indicators, lag periods, mathematical functions, and rule combinations, which produce good results while passing a plausibility test? It is possible to use PSO for the same combinatorial task of combining good rule fragments, where the rule is encoded as a real-valued vector. Initially a population of particles (random vectors) is created, each corresponding to a trading rule. The fitness of each particle or rule can be tested using historical financial data, and the particle swarm algorithm is iteratively applied to uncover better trading rules.

Alternatively, a social programming particle swarm algorithm in the form of grammatical swarm could be used in a manner similar to grammatical evolution to generate rules for a trading system (see Chap. 4 and Part III).

5.7 Recent Developments in PSO

There is a growing literature on particle swarm algorithms that extends the underlying algorithm presented in this chapter. As a starting point the interested reader should refer to work on speciation of particles with a predator-prey metaphor [198] and other modifications to the behaviour of the basic algorithm, e.g., [21, 22, 154]; modelling the adaptation of organisations on a strategic landscape with OrgSwarm [33, 34]; and, as outlined earlier in this chapter, the generation of programs using grammatical swarm [168].

5.8 Summary

The key learning mechanisms in the PSO algorithm are driven by a metaphor of social behaviour: that good solutions uncovered by one member of a population are observed and copied by other members of the population. Of course, these learning mechanisms abound in business and other social settings. Good business strategies, good product designs, and good theories stimulate imitation and subsequent adaptation. Particle swarm algorithms have proven to be successful optimisation tools in a variety of applications, and they have clear potential for application to financial modelling.

Ant Colony Models

Ant colony models (ACM) constitute a family of population-based, optimisation and clustering algorithms that are metaphorically based on the activities of social insects. They are inspired by the observation that social insects such as ants, bees and termites live in highly organised colonies. Despite the high degree of organisation of these colonies, there is no overt top-down communication structure. Each individual insect follows a fairly limited set of rules, usually with only local awareness of its environment, but the interaction of the activities of these individuals gives rise to a complex *emergent*, self-organised structure, and provides the colony with the ability to adapt to alterations in its environment. ACM emphasises the importance of local communication (or *distributed learning*) between the individuals in a population in permitting the population to adapt successfully over time.

The information-exchange mechanism between ants can be direct (by visual or chemical contact) or indirect (where an ant modifies the environment faced by its peers). In the latter case, the actions of individual ants have the effect of changing the way the environment, or a problem, is perceived by other ants. This form of communication is known as *stigmergy*.

In general, ant algorithms are derived from four metaphors of ant behaviour (Fig. 6.1). In this Chapter we limit attention to ant-foraging models which can be used for discrete optimisation problems. A more detailed discussion of ACM can be found in [23, 64, 66].

6.1 Ant-Foraging Models

One way that ants can alter the environment faced by their peers is by marking trails to discovered food sources. During ant food-foraging, individual ants lay down a chemical trail of pheromone. If a group of ants search randomly around their nest for food, pheromone trails from the nest to close-by food sources will tend to grow in strength more quickly than those to far-away food sources, as ants travelling to the closest food source will return quickly to the

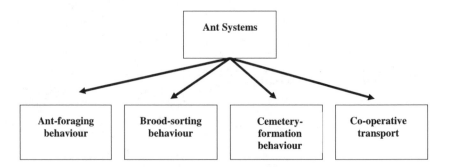

Fig. 6.1. Taxonomy of ant colony algorithms

nest, leaving an outward and inward trail of pheromone. The quality of the food source may also affect the amount of pheromone deposited. Typically, the better the food source, the more pheromone deposited. If subsequent foraging ants have a tendency to follow stronger rather than weaker pheromone trails, *auto-catalytic* behaviour will emerge, with ever-increasing numbers of ants travelling along the strong trail and reinforcing it further [65] (Fig. 6.2). This gives rise to a positive feedback loop between the ants in the search for food. The effect of the pheromone-following behaviour is to create an indirect communication between ants. As trails emerge over time a collective *memory* is created as to the location of the food source.

One drawback of the positive feedback mechanism is that it can lead to *lock-in*, whereby the heavily pheromone-reinforced path continues to be used, even if a rich food source subsequently becomes available closer to the nest. The concepts of positive feedback and lock-in are not unique to ant societies. For example, telecommunication and transport networks in human societies display similar characteristics, whereby once a given technology becomes dominant it becomes difficult to displace.

6.1.1 Ant-Foraging Algorithm

Ant-foraging behaviour can be used to design an algorithm for discrete optimisation. The basic flowchart of an ant-foraging algorithm is outlined in Fig. 6.3. In the algorithm, a colony of artificial ants iteratively and stochastically constructs solutions (a construction graph) for the problem of interest, using artificial pheromone trails, which are modified as the algorithm runs. During the solution construction stage, each ant builds a complete solution to the problem by combining discrete solution fragments. The nature of these fragments depends on the problem. For example, in a travelling salesman problem (TSP) the fragments will correspond to the choice of the next city to be visited on the tour being constructed. The choice of solution fragments by individual ants at each step of the construction routine is guided by the

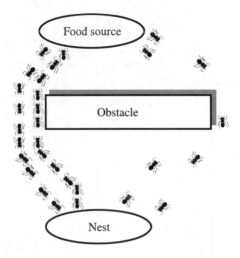

Fig. 6.2. Foraging ants reinforcing a trail to a food source

quantity of pheromone deposited on each of the possible choices they face, with higher pheromone choices being more likely to be selected by an ant. After all the ants have constructed a solution to the problem, the pheromone deposits on each solution fragment are updated, with solution fragments on higher-quality solutions receiving stronger reinforcement. Over multiple iterations of the algorithm, better solution fragments become more heavily used by the population of ants, and less-successful fragments fall into disuse. Typically, the pheromone trails are subject to an *evaporation* process during the update step, to ensure that less-travelled solution fragments are forgotten over time by the population of ants.

Table 6.1. Correspondence between ant systems and optimisation

Ant System Component	Correspondence
Complete ant trail	Solution: rule, equation or program.
Fragment of ant trail	Fragment of solution.
Pheromone laid down on ant trail.	Memory of the quality of past solutions.
Updating of pheromone trails and probabilistic choice of rule fragments from alternatives available.	Creating diversity in search process.

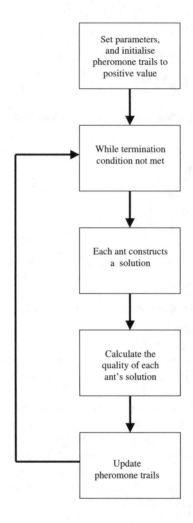

Fig. 6.3. A flowchart of the ant-foraging algorithm

Applying the Algorithm

Ant colony optimisation (ACO) uses metaphorical inspiration from ant be-haviours to create algorithms for optimisation purposes. In applying the ant-foraging algorithm to solve a combinatorial optimisation problem, two key questions must be addressed by the modeller.

   i. How do the ants construct proto-solutions to the problem of interest?
  ii. How are the pheromone trails updated?

At the start of the algorithm, all the potential solution fragments have their pheromone initialised to a non-zero value. Each ant in turn commences a traversal of the potential solution fragments, in order to build a complete solution. There are many possible ways to implement the construction process. At each decision juncture when a solution is being constructed an ant could simply select from the set of available solution fragments the fragment which has the highest pheromone level (this corresponds to a greedy search process). However, this would tend to result in rapid convergence to a small set of solutions. Another possibility is to stochastically choose amongst the discrete solution fragments available at each step of the construction process. For example, the probability of choosing fragment i from amongst the K possible choices at a particular construction step could be determined using (τ_k is the quantity of pheromone associated with fragment k):

$$Prob_i = \frac{\tau_i}{\sum_{k=1}^{K} \tau_k} \tag{6.1}$$

This approach ensures that while solution fragments which have been part of good solutions in the past are more likely to be selected as their pheromone levels are high, an ant still has the potential to explore a new path. A more complex approach is to combine the pheromone information with an estimate of the likely quality of each of the solution fragments when making the choice of which fragment to add to the growing solution. This is known as adding *visibility* to the construction process (see [23] for more information on this idea).

After all the ants have traversed the solution fragments and have constructed solutions to the problem, the quality of each of these solutions is assessed, and this information is used to update the pheromone trails. The update process typically consists of an *evaporation* step, and a pheromone deposit step:

$$\tau_i(t+1) = \tau_i(t)(1-p) + \delta_i \tag{6.2}$$

In the evaporation step the pheromone associated with every solution fragment is degraded, where the evaporation rate is given by p. The evaporation rate crucially controls the balance between exploration and exploitation in the algorithm. If p is close to 1, then the pheromone values used in the next iteration of the algorithm are highly dependent on the good solutions from the current iteration, leading to local search around those solutions. Smaller values of p allow solutions from earlier iterations of the algorithm to influence the search process.

The amount of pheromone deposited on each solution fragment i (δ_i) during the pheromone update process depends on how the deposit step is operationalised in the algorithm. The deposit step can be performed in many ways, depending on which solutions are chosen to participate in the deposit process, what weight is applied to each of these solutions in the deposit process, and how pheromone is deposited on each fragment of a solution participating in the deposit process.

In choosing which solutions participate, one method would be to select only the best solution in the current population. Another method would be to allow all solutions in the current population to play a role in the deposit process. Elitist strategies which allow the best solution found so far over all iterations of the algorithm to participate in the deposit process can also be implemented.

The updating of the pheromone deposits on individual fragments of solutions can also be performed in a variety of ways. One method is to reinforce the components in the solution found by each ant (if all ants participate in the update process) by adding $Q * F$ to the pheromone associated with each solution fragment, where Q is a modeller-chosen fixed amount of pheromone, and F is a measure of the quality of the solution scaled into the range $(0 \rightarrow 1)$. Therefore, solution fragments contained in good-quality solutions are more heavily reinforced by pheromone.

To avoid premature convergence of the search process to a single solution, and to ensure that solution fragments have a non-zero chance of being selected at each step of the construction process, the pheromone associated with each solution fragment (τ_i) may be constrained so that after the pheromone update process $0 < \tau_{min} \leq \tau_i \leq \tau_{max}$. These bounds prevent the relative differences between the pheromone trails associated with each fragment from becoming too extreme. The algorithm is terminated either after a fixed number of iterations, or when there has been no improvement in the best solution for a set number of iterations.

6.2 A Financial Application of ACO

In applying ACO to real-world problems, the trails formed by individual ants are potential solutions (for example trading rules) for the problem of interest. Individual elements of these trails represent choices for fragments of each solution hypothesis. Hence, the individual ants traverse choices for rule fragments, and the entire trail traversed by an ant represents a complete solution. The better the solution in terms of solving the problem, the greater the strength of the pheromone trail attributed to it, and the greater the chance that subsequent ants will select to travel along it. Let's take the case of constructing a simple IF-THEN trading rule:

$$IF\ [INDICATOR_1(time)\ (<,>,=)\ VALUE_1]$$

$$THEN\ (BUY,\ SELL,\ DO\ NOTHING)$$

If we restrict attention to the case where there is a finite number of choices for each of the components of the rule, the selection of a good rule represents a combinatorial optimisation problem, in that there are multiple possible choices for the 'IF condition', the choice of indicator, the choice of time lag and so on.

In applying an ant-foraging algorithm to develop good trading rules, each ant in the population initially starts with an empty rule, and adds one term/parameter at a time to the rule from a set of possible choices for each term/parameter, until a full trading rule is developed. The choice of term to be added at each stage is biased based on the quantity of pheromone associated with each of the possible choices. On the first iteration of the algorithm, each choice is randomly initialised with a small positive value of pheromone.

The quality of the trading rule developed by each ant can be evaluated by testing it against a historical financial dataset. Once all ants have constructed rules (of differing quality), the pheromone associated with each choice of trading rule and its components is updated. Better rules attract more pheromone. To reduce the chance that the process will converge to a single solution too quickly, the quantity of pheromone associated with all trails is subject to an evaporation or forgetting process after each iteration of the algorithm. Pheromone trails corresponding to better trading rules tend to get strengthened (reinforced) over time as more ants traverse them. This leads to more-intensive searching of variants of the best-so-far trading rules. The process is iterated until stopped by the modeller, with the strongest trails corresponding to the best trading rules. Of course, ACO is a general-purpose optimising algorithm, and its utility is not limited to the construction of trading rules.

6.3 Ant-Inspired Classification Algorithms

Another behaviour of ant colonies which has been used as a metaphor to create classification and clustering algorithms is the picking up and depositing of items into clusters of like items. Examples of this behaviour include *brood-sorting*, where ant larvae are sorted and grouped into piles of differing size, and *cemetery building*, where dead ants are removed from the colony and deposited together.

Typically in these algorithms, vectors of information are sorted into similar groups. An illustration of how this approach can be used to create a model for predicting corporate failure is provided in a case study in Chap. 18.

6.4 Hybrid Ant Models

Like other biologically inspired computational methodologies, ant models need not be applied on a stand-alone basis, and can be employed in a hybrid fashion. One example of this is the creation of a *MLP-ACO* hybrid, where an ant algorithm is used to train the weights in an MLP structure [207]. An ACO algorithm could also be used as a search engine in a GE system.

6.5 Summary

Ant colony systems comprise of a family of algorithms which draw their metaphorical inspiration from the activities and learning mechanisms of social insect societies. A key feature of these societies is their ability to promote problem-solving behaviour between individuals in the absence of a top-down control system. The algorithms can be used for both classification and optimisation purposes.

7

Artificial Immune Systems

In previous chapters, we saw that a variety of biologically inspired algorithms can be used for classification and optimisation purposes. In this chapter, a number of *artificial immune system* (AIS) algorithms, whose design is inspired by the workings of the immune system, are outlined. The object in designing and applying AIS is not to produce exact models of the natural immune system, rather the aim is to extract ideas and metaphors from the workings of the natural immune system which can be used to help solve real-world problems. The most commonly applied AIS algorithms can be grouped into three categories (Fig. 7.1), based on distinct features of the natural immune system.

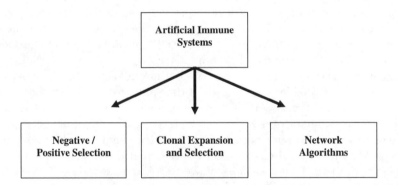

Fig. 7.1. Taxonomy of AIS algorithms

In this chapter we will primarily concentrate on the negative selection algorithm which can be used for classification, and the clonal expansion and selection algorithm which can be used for optimisation. Before we examine these algorithms, a short overview of the workings of the natural immune system is provided. Readers requiring a detailed introduction to the immune system are referred to [93] and [116].

7.1 Overview of Natural Immune Systems

The immune system is comprised of an intricate network of specialised tissues, organs, cells and chemical molecules. The capabilities of the natural immune system include the ability to recognise, destroy and remember an almost unlimited numbers of pathogens (foreign or *non-self* objects that enter the body, including viruses, bacteria, multi-cellular parasites, and fungi), and also to protect the organism from misbehaving cells in the body. To assist in protecting the organism, the immune system has the capability to distinguish between *self* and *non-self*. Critically, the system does not require exhaustive training with negative (non-self) examples to make these distinctions, but can identify a pathogen as being non-self even though it has never been encountered before.

7.1.1 Innate vs Adaptive Immunity

The human natural immune system has multiple-level defenses (Fig. 7.2). The first lines of defence are barriers which physically block ingestion of pathogens such as skin and nasal hair. These barriers are supported by physiological defenses, fluids secreted by the body (saliva, sweat and tears), which move pathogens out of the body and/or contain disruptive enzymes. In addition, humans have both an *innate* (or non-specific) and an *adaptive* (or specific) immune system [18]. The innate immune system uses a number of reliable signatures of non-self, such as pathogen-associated molecular patterns, to identify pathogens. An example of such a pattern is the mannose carbohydrate molecule which is found in many bacteria but not in mammals. These patterns have remained stable for long periods of time and are encoded in the genome of our immune systems [38]. The innate immune system is present at birth and it does not adapt over a person's lifetime. If the innate immune system cannot remove an invading pathogen, the adaptive immune system takes over. Adaptive immunity is directed against specific pathogens and is modified by exposure to them. Thus a *memory* of previous invaders, and how to deal with them, is created and maintained by the immune system.

7.1.2 Components of the Immune System

Both the innate and acquired immune systems are comprised of a variety of molecules, cells and tissues. The most important cells are leukocytes (white blood cells) which can be divided into two major categories: phagocytes, and

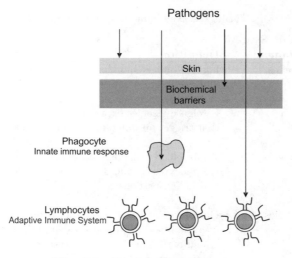

Fig. 7.2. Natural immune system

lymphocytes. The first group belongs to the innate immune system while the latter group mediates adaptive immunity. In this review we concentrate on the adaptive immune system.

Lymphocytes circulate constantly through the blood, lymph, lymphoid organs and tissue spaces. A major component of the population of lymphocytes is made up of B and T cells. These cells are capable of recognising and responding to certain antigen (foreign molecules) patterns presented on the surface of pathogens. Antigens are not the invading pathogens themselves, rather they are molecular signatures expressed by the invading pathogen. A major role in this recognition process is played by molecules of the *major histocompatibility complex* (MHC) [107]. These molecules act to transport peptides (fragments of protein chains) from the interior regions of a cell and present these peptides on the cell's surface. This mechanism enables components of the immune system to detect infections inside cells, without having to penetrate the cell's membrane.

The control of adaptive immunity can be divided into two branches: *humoral immunity* which is controlled by B cells, and *cellular immunity* which is controlled by T cells. Humoral immunity is mediated by specially designed proteins or *antibodies* contained in bodily fluids (or 'humors'), and it involves the interaction of B cells with antigens. Cellular immunity is cell-mediated, and plays an important role in the killing of virus-infected cells and tumours.

B Cells and T Cells

B cells and T cells have receptors on their surfaces which are capable of recognising antigens via a binding mechanism. The surface of a B cell contains

Table 7.1. Key immune system terms

Immune System Component	Definition
Pathogens	Foreign bodies including viruses, bacteria, multi-cellular parasites, and fungi.
Antigens	Foreign molecules expressed by a pathogen that trigger an immune system response.
Leukocytes	White blood cells, including phagocytes and lymphocytes (B and T cells) for identifying and killing pathogens.
Antibodies	Glycoproteins (protein+carbohydrate) secreted into the blood in response to an antigenic stimulus that neutralise the antigen by binding specifically to it.

Y-shaped receptors (or antibodies). Antibodies possess two *paratopes* (each arm of the Y-shaped receptor), which are used to match or identify molecules. These molecules may represent proteins or fragments of proteins making up an antigen on the surface of a pathogen. The regions on the antigen that a paratope can attach to are called the *epitopes*. Identification of the antigen is achieved by a complementary matching between the paratope of the antibody and the epitope of the antigen. The match between the paratope and epitope need not be perfect. To increase the number of pathogens that the immune system can detect, individual lymphocytes can bind to a variety of antigens. This enhances the power of the immune system, as multiple lymphocytes will bind to an invading pathogens therefore there will be multiple signals created in the immune system that an invader has been detected. The closer the match between paratope and epitope, the stronger the molecular binding between the antibody and the antigen, and the greater the degree of *stimulation* of the B cell.

T Cell-Dependent Humoral Immune Response

When an antibody of a B cell binds to an antigen, the B cell becomes stimulated. The level of stimulation depends on the closeness or *affinity* of the match between the antibody and the antigen. Once a threshold level of stimulation is reached, the B cell is *activated*. Before activation takes place, the B cell must be *co-stimulated* by a variant of the T cell population called *helper T cells*. When helper T cells recognise and bind to an antigen, they secrete cytokines, which are soluble proteins that are used to provide signaling between the cells of the immune system. In addition to the cell-cell interaction where the T cell can bind to a B cell, the secreted cytokines can act on B cells to co-stimulate them.

Once the stimulation level of a B cell has reached a threshold level, the B cell is transformed into a *blast cell* and completes its maturation in the lymph nodes where a *clonal expansion* and *affinity maturation* process occurs. The object of the clonal expansion process is to generate a large population of antibody secreting cells and memory B cells which are specific to the antigen. In the lymph nodes, activated B cells begin to clone at a rate proportional to their affinity to the antigen that stimulated them. These clones undergo a process of affinity maturation in order to better tune B cells to the antigen which initiated the immune system response. When new B cells are generated, the DNA strings that encode their receptors are subject to recombination, mutation and insertion processes, and new forms of receptors are constantly created. When a tailor-made detector is required for a specific novel antigen, the ability of the immune system to generate diversity is enhanced by means of a high mutation rate in the cloning process, for the genes which encode the B cell's Y-shaped receptors (this process is known as *somatic hypermutation*), and the differential selection of the clones which best match the antigen. The evolutionary process of creating diversity and the subsequent selection of the variants of lymphocyte that become increasingly specific to an antigen is referred to as *clonal selection*.

The T cell-dependent humoral immune response is a series of complex immunological events. It commences with the interaction of B cells with antigens. The B cells which bind to the antigen are co-stimulated by helper T cells, leading to proliferation and differentiation of the B cells to create B plasma and memory cells. The new plasma B cells secrete antibodies (immunoglobulins) which circulate in the organism and mark the antigens by binding to them. These antigens and the associated pathogen are then targeted by the immune system for destruction. The steps in the process can be summarised as follows:

i. Antigen-secreting pathogen enters the body.
ii. B cells are activated by the foreign antigen.
iii. With help of T cells, B cells undergo cloning and mutation.
iv. Plasma B cells secrete immunoglobulins which attach to the antigen.
v. Marked antigens are attacked by the immune system.
vi. Memory of the antigen is maintained by B memory cells.

T Cell Tolerogenesis

A major challenge for the immune system is to ensure that only foreign or misbehaving-self cells are targeted for destruction. The system must be able to differentiate between self (proteins and molecules which are native to the organism) and non-self (cells and molecules which are foreign). In the normal creation of T cells, their receptors are randomly generated, and so could potentially bind to either self or non-self. To avoid auto-immune reactions where the immune system attacks its host organism, it is theorised that the cells must be self-tolerised. In the case of T cells, this process of tolerogenesis takes place in the thymus. One mechanism for conferring self-tolerance to the lymphocytes as they are maturing is exposure to a series of self-proteins. Any lymphocyte that binds to self-proteins is killed, and only self-tolerised cells are allowed into the circulation system for lymphocytes. This represents a negative selection process as only non-self reactive T cells are permitted to survive.

Immune System Memory

If the immune system encounters an antigen for the first time, a *primary response* is provoked in the adaptive immune system, and the organism will experience an infection while the immune system learns to recognise the antigen. In response to the invasion, a large number of antibodies will eventually be produced by the immune system which will help eliminate the associated pathogen from the body. After the infection is cleared, a memory of the successful receptors is maintained to allow much quicker *secondary response* when the same or similar pathogens invade thereafter. The secondary response is characterised by a much more rapid and more abundant production of the relevant antibody than the primary response. If a close, but not identical, variant of the pathogen is later encountered, a secondary response can be provoked by an antibody to the original antigen which is a sufficiently close match for the differentiated antigens on the new pathogen. Therefore if a mutated version of the original pathogen is encountered, the immune system is already partly adapted to deal with it, based on its previous learning. This is the concept underlying the process of immunisation against a disease using a non-harmful variant of that disease. Although there is debate as to the precise nature of how immune system memory is maintained, in broad terms the immune system maintains a population of long-lived lymphocytes or *memory cells*. Both T and B cells have memory variants. The creation of memory cells ensures that the results of past learning are physically encoded into the current population of lymphocytes.

Danger Theory

Although the concept of self vs non-self provides a fertile ground for the development of algorithms for anomaly detection, a variety of alternative views of the working of the immune system exist. The *Danger Theory*, proposed by Matzinger [149, 150], challenges the traditional self vs non-self view of the immune system, and although the theory is not complete it has attracted the interest of immunologists in the past decade [1]. Danger theory suggests that the self/non-self distinction is not sufficient to explain immune system behaviour as not all foreign bodies are reacted to by the immune system (for example, consuming food or breathing air does not provoke an immediate immune system response), and some self-cells (for example, faulty cells) are targeted by the immune system. Matzinger notes that the human body changes over its lifetime, hence there is no static notion of self. The central tenet of the Danger Theory is that the immune system does not react naively to non-self, but rather reacts to danger. Under this theory, it is considered that a cell in distress sends out an alarm signal, whereupon antigens in the neighbourhood of this cell are captured by antigen-presenting cells such as macrophages. The danger signal creates a zone of immune system activity around its place of origin, and B cells producing antibodies that match antigens within the danger zone get stimulated and undergo the clonal expansion process. The key idea in danger theory is that the immune system focusses on events which trigger a danger signal, and does not therefore react to harmless non-self.

7.2 Designing Artificial Immune Algorithms

Even from the brief description of the natural immune system, it is apparent that the system is intricate and complex. From a modelling perspective, it can be considered as a sophisticated information processing system which possesses powerful pattern recognition and classification abilities. It also has the capability to adapt to new circumstances (problems), and can remember solutions to problems it has previously encountered.

In designing artificial immune algorithms (AIAs) the object is to draw metaphorical inspiration from the workings of the natural immune system, or theories of the workings of this system, to design algorithms which can be applied to solve computational problems. AIA typically use a limited number of components, often only a single lymphocyte and a single form of antigen. An application will require that both a similarity/affinity measure, and a fitness measure are defined. Although a multitude of metaphors could be drawn from natural immune systems for the purposes of designing AIA, we will focus on two: the negative selection algorithm, and the clonal expansion and selection algorithm.

7.2.1 Negative Selection Algorithm

The basis of the negative selection algorithm is the ability of the immune system to discriminate between self and non-self, or more broadly to distinguish between two system states, normal or abnormal. Forrest et al. (1994) [80] developed a negative selection algorithm analogous to the negative selection or self-tolerogenesis process during T cell maturation in the thymus. Initially a predetermined number of detectors are created randomly. During the training (tolerogenesis) process any detector that falls within a threshold distance r_s (usually measured using Euclidean distance) of any elements of the set of self samples is discarded and replaced with another, randomly generated detector. The replacement detector is also checked against the set of self samples. The process of detector generation is iterated until the required number of valid detectors is generated (Fig. 7.3). All of the resulting detectors are potentially useful detectors of non-self. The pseudo-code for the algorithm is as follows (S is the set of self-samples, r_s is a predefined self-radius, and it is assumed that the search-space is bounded by an n-dimensional (0,1) hypercube):

i. Detector set (D) is empty
ii. Repeat
iii. Create a random vector $\mathbf{x}$, drawn from $[0, 1]^n$
iv. For every $\mathbf{s_i}$ in $S, \mathbf{s_i} : i = 1, 2, ..., m$
v. Calculate the Euclidean distance (d) between $\mathbf{s_i}$ and $\mathbf{x}$
vi. If $d \leq r_s$ go to step (ii)
vii. Add $\mathbf{x}$ (a valid non-self detector) to set D
viii. Until D contains the required number (assume $\mathbf{N}$) of valid detectors

Once a population of detectors has been created, they can be used to classify new data observations. To do this, the new data vector is presented to the population of detectors, and if it does not fall within r_s of any of them, the data vector is deemed to be non-self. Otherwise, the new data vector is deemed to be self. A crucial point in the negative selection process is that the immune system does not require specific examples of non-self in creating its detectors. Potentially, the detectors can uncover any instance of non-self, even those never before encountered.

In using the negative selection algorithm, a choice must be made as to the value of self-radius r_s, and the number of detectors to use. The choice of value for r_s seeks to balance the detection rate and the false-alarm rate of the system. If a small value of r_s is used the detection rate for non-self will be low, and if a high value of r_s is set the false alarm rate will be high.

7.2.2 Clonal Expansion and Selection Algorithm

This algorithm is inspired by the clonal selection and affinity maturation process of B cells once the immune system has detected a pathogen. The object of the clonal selection process is to create a large quantity of antibodies which will bind strongly to a specific antigen.

Adopting this metaphor in order to design an optimisation algorithm, the antibody can be considered as a potential solution, the antigen is a test dataset, and the degree of the binding or fit between the antibody and the antigen represents the fitness or the quality of the solution. The objective, therefore, is to start from an initial population of solutions, test them against the dataset, and, using the algorithm iteratively, improve the quality of the solutions in the population. The clonal selection metaphor can be turned into an optimisation algorithm in a variety of ways. An outline of an algorithm based on the CLONALG algorithm [56] is:

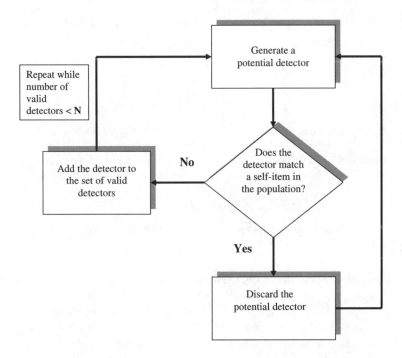

Fig. 7.3. A flowchart of the creation/training process for detectors in the canonical negative selection algorithm

   i. Create an initial random population P of solution vectors (antibodies).
  ii. Select a subset F of the solutions from P, biasing the selection process towards better solutions.
 iii. For each member of F (the parents), create a set of clones, with better members of F producing more clones.
 iv. Mutate each of these clones, in inverse proportion to their parent's fitness (the hypermutation step). Better solutions are mutated less.
  v. Select a subset of the newly generated solutions S.
 vi. Create a number of newly created random solutions R.
 vii. Replace poorer members of P with better solutions from S and R.
viii. Repeat steps (ii)-(vii) until a terminating condition is triggered.

Clonal selection algorithms can therefore be used for optimisation purposes, just like evolutionary algorithms or ant algorithms. The key difference between each of the groups of algorithms is the method they use for generating variety when seeking to iteratively improve solutions.

Table 7.2. Correspondence between clonal selection and optimisation

Immune System Component	Correspondence
Antibody	Rule, equation or program.
Antigens	Test data.
Antibody-antigen matching	Quality of the rule or program.
Cloning and mutation	Generation of variety in order to uncover better solutions.

7.3 Financial Application of the Negative Selection Algorithm

One financial application of the negative selection algorithm would be to create a classification system to predict whether a company will fail in the near future. Assume that the diagnosis is to be made on the basis of ratio information drawn from the financial statements of companies, and that a dataset has been collected which provides examples of ratios for financially healthy companies and also for companies which later went bankrupt.

Self is defined as financially healthy companies. Next a set of detectors (of size D) is randomly created. Each of these detectors consists of a vector of real numbers, corresponding to a set of accounting ratios. The negative selection process is then applied, whereby a series of vectors of ratios corresponding to the healthy companies is presented to the detectors. Any detector which is identical or similar (the degree of similarity could be measured using Euclidean distance) to a data vector corresponding to a healthy company is discarded. As detectors are discarded, new detectors are randomly created to build up the size of the population to D, and the newly created detectors are subject to the same negative selection process. If a detector is created which fails to match any vector in the training set of healthy companies, it is a potentially useful detector of an unhealthy or failing company. Once a population of detectors is generated, the detectors can be exposed to new data, and used to predict whether these companies will fail or not. New data vectors which trigger a detector, and which therefore have characteristics similar to failing companies, are classed as failing. Otherwise the new vector is classed as a financially healthy company.

Figure 7.4 provides a graphical representation of the model at the end of the training process, where the area covered by detectors corresponds to an unhealthy zone of financial ratios. For ease of display, only two ratios are considered, and both have been normalised into the range (0,1). Any companies outside the zone of the detectors is classed as a healthy company.

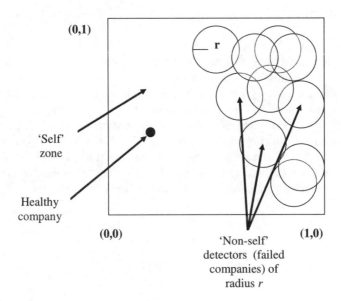

Fig. 7.4. Example of location of detectors for corporate failure model after training. Each axis corresponds to a financial ratio

While this basic algorithm will produce an immune-inspired classification system, it will not be particularly efficient for a number of reasons [39]:

i. It ignores the fact that examples of past 'failing companies' exist.
ii. The task of generating a population of valid detectors will grow rapidly as the size of self increases.

The first issue can be overcome by using historical examples of accounting ratios from failing companies to seed the process of creating valid detectors, thereby speeding up the process of creating valid detectors. The impact of the second issue can be lessened by altering the algorithm to permit the creation of variable-size detectors [117]. This could allow the coverage of large areas of self space with a number of large detectors.

The negative selection algorithm can be applied to a wide variety of settings. Self can be broadly defined as a normal pattern of activity, or as stable behaviour of the system/process of interest. Dasgupta and Forrest [50] provide an example of how the negative selection algorithm could be applied to detect novelty (defined as a change in the steady-state characteristics or normal behaviour of the system), in time-series data. Generalising this idea to the case of designing trading systems, an AIA could be constructed to identify different market *regimes*, and supply a feed into a trading system which then selects a trading strategy which is suitable for current market conditions.

7.4 Summary

Although natural immune systems are complex and consist of a huge number of individual components, they can be considered as a distributed, self-organising system which operates in a dynamic environment. The mechanisms of natural immune systems, including their ability to distinguish between self and non-self states, and their ability to maintain a memory of previous invaders, provide a rich metaphorical inspiration for the design of pattern-recognition and optimisation algorithms. In this chapter we have discussed how two of these metaphors, the negative selection process for T cells, and the clonal selection and expansion of B cells, can be applied to create a classification system and an optimisation algorithm, respectively.

Part II

Model Development

8

Model Development Process

Part I of this book described a variety of biologically inspired algorithms, and described how these could be used for modelling purposes. Part II discusses the process of actually developing a high-quality financial model, concentrating on the development of a market trading system.

The construction of any financial model is a multi-stage process, consisting of the determination of the goals for the project followed by data collection, data preprocessing, model construction, data postprocessing, model validation, and finally model implementation. It cannot be overemphasised that the degree of success of any model/trading system is critically impacted by the rigour of all the steps in its development process, not just the sophistication of the biologically inspired algorithm(s) embedded in it.

8.1 Project Goals

The object of a trading system is to act as a screening mechanism, to decide which financial assets to buy (or sell), and when to buy (or sell) them. At the heart of any trading system is a predictive model for the market which is being traded. Having selected the market of interest, the modeller must define:

- what the model will forecast; and
- what performance measure is appropriate for the model.

8.1.1 What to Forecast?

Although it may seem obvious that the common goal in financial prediction is to forecast the future price of an asset, this is not necessarily the case, and, in any event, a raw price will not usually be an easy predictive target. Suppose the S&P 500 index is currently 1200, and that it changes by an average of 10 points per day. If the objective is to forecast the next day's

Fig. 8.1. Seven steps in building a trading system

closing value of the index to within 3 points, this corresponds to a required accuracy of approximately 0.25% (3 out of 1200), presenting a difficult task for any predictive model. If the goal of the model is altered to predict the one-day *change* in the index instead of the absolute level of the index, the required accuracy drops to 3 out of 10. Changing the predictive target can substantially impact on the ease of the predictive task. Other predictive targets which can be used apart from the magnitude of change in price over the next x days include the direction of the change in price ($+$/-) in the next x days, whether a (for example) 2% move in price will occur in the next x days, or the expected market condition (trending up, down or non-trending) over the next x days. Accurate predictions of any of these would provide a very useful input into a trading system.

8.1.2 What Performance Measure Is Appropriate?

Although many algorithms (including canonical feedforward MLPs) minimise error measures such as mean squared error (MSE) or root mean squared error (RMS), models constructed using these criterion may not perform well when used for trading purposes. *Squared error* based goodness-of-fit criteria will tend to penalise large errors heavily during the model construction process, but this does not guarantee that the final model will be good at correctly identifying and avoiding all large errors. MSE could be low, not because the model makes no large errors, but perhaps because the model is accurate for small changes in price which cannot be profitably traded due to trading costs, while the model misses a substantial number of larger changes which could have been traded profitably if they had been anticipated. Similar problems can arise with the use of an R^2 goodness-of-fit criterion, based on the correlation of the model's prediction and actual market changes.

Metrics such as MSE or R^2 assume that the costs of predictive errors are symmetric. This is not the case in financial prediction. The cost of an error to a trader depends on both its magnitude and direction. If a model generates a buy signal, but underestimates the size of the upward price movement, the trader makes a profit despite the prediction error. A numerical prediction error of similar magnitude but in the other direction (the model predicts that a share's price will increase but in fact it decreases) means the trader makes a loss. This problem can be alleviated by using an *asymmetric error function* to more heavily penalise errors where the direction of the predicted price change is wrong.

The choice of error measure depends on the use to which the model will be put. If the model is to be used for trading purposes, the most appropriate measures of performance are trading returns, scaled by a measure of risk. The trading characteristics of the developed system will depend critically on how returns and risk are defined. The risk associated with a trading system can be measured in a large number of ways, including its *drawdown*, which is the maximum cumulative trading loss of a system during its training or testing period. Hence a performance metric such as the *Stirling ratio*:

$$\frac{\text{Return}}{\text{Drawdown}} \tag{8.1}$$

could be used to evaluate a trading rule. A variant on this is the *modified Stirling ratio* [61]:

$$\frac{\text{Return}}{1+\text{Modified drawdown}} \qquad (8.2)$$

where the modified drawdown is defined as being the max(drawdown, or 2% of the current position). One advantage of the modified form of the ratio is that it is more robust to minor changes in the value of drawdown, when the absolute size of the drawdown is small. Another common risk measure is the *Sharpe ratio* which compares the level of excess returns (defined as the trading returns less the risk-free returns generated by a trading system over a period of time) with the volatility of those returns:

$$\frac{\text{Trading profit - Risk free return}}{\text{Standard deviation of trading profit}} \qquad (8.3)$$

As for the Stirling rations, high values of the Sharpe ratio are preferred. The choice of performance (or fitness) function will determine how often the system trades and what percentage of its trades are winning trades. For example, a trading system could be biased to:

i. maximise the ratio of average trade profit to maximum drawdown,
ii. maximise the Sharpe ratio, or
iii. minimise the volatility of trading returns.

A particular advantage of using a methodology such as GE to construct a trading system is that there are no requirements that the performance measure is differentiable. Hence, the evolution of trading systems can be biased towards whatever risk/return relationship is preferred by the trader. The evolutionary process could be biased to favour trading rules which produce good returns with low drawdown, or a constraint could be placed that only trading rules which produce a drawdown of less than $x during the training period will be permitted to evolve.

8.2 Data Collection

No matter how sophisticated the biologically inspired algorithm, the old adage *garbage-in-garbage-out* (GIGO) applies. The success of a modelling effort is largely determined by the quality of the data collected, its preprocessing, and the postprocessing of the resulting outputs.

8.2.1 Trading Philosophy

Before detailed consideration can be given to selection of variables for inclusion in a trading system, the *trading intent* and *trading time-horizon* of the system must be defined. The trading intent arises from the strategy the trader intends to adopt in identifying which financial assets to buy and sell, and must be based on an underlying hypothesis as to how the market behaves (see Chap. 15) for an example of how trading patterns may be influenced by the time horizon adopted by an investor). In essence, the process of constructing a trading system consists of formulating a hypothesis of how the market works, and then testing whether the hypothesis holds up when tested on real data.

The selection of potentially useful explanatory variables is also impacted by the intended trading time-horizon. If a model is being constructed with a view to assisting with long-term equity investment decisions, the relevant variables will be those which help assess the long-term prospects for a firm's shares, and the investor may focus on variables which may indicate that the value of a share has moved out of alignment with its fundamental value. The trading intent is therefore to identify under-(or over) valued shares, and hold the trading position for a period of time. Therefore, minor intra-day movements in share price will not be important.

If attention is placed on short-term trading, short-term price changes will be relevant. Is the intent to hold positions for a few days, or is the intention to build a system for intra-day trading? For either of these time-horizons, the focus will shift from long-term indicators of value, which will be invariant in the short term, to daily flows of demand and supply in the market. Therefore, the relevant explanatory variables may include technical indicators of the underlying forces of supply and demand for the asset and inter-market data concerning the values of related financial assets in other markets.

Whatever the trading intent and horizon, there must be a plausible relationship between the selected inputs and the predictive target. The trader cannot simply throw a group of input variables into a biologically inspired search engine and expect it to automatically uncover something interesting. Throwing 'everything' in and hoping something useful comes out will most likely result in the production of a spurious model.

What Variables?

The range of potentially useful variables will vary depending on the market the trading system is being constructed for. Is the model intended to trade equities, foreign exchange, commodities or financial derivative products? Taking equity markets as an example, three primary sources of information exist:

- technical indicators,
- intermarket indicators, and
- fundamental indicators.

It is not possible to give a complete discussion of each of these sources of information in a single text and only a brief introduction to them is provided. Technical indicators are explanatory variables formed from various combinations of current and historic asset price and transaction volume information. They are widely used in short-term trading systems, and the language of technical analysis permeates the financial press. Technical indicators are underpinned by the concept of technical analysis. Advocates of technical analysis consider that it can be used to preprocess historic price/volume information to uncover patterns, which when they recur can be recognised and traded on. The next chapter discusses technical analysis, and hence only intermarket and fundamental indicators are considered here.

Intermarket Indicators

Analysis of intermarket indicators attempts to highlight when divergences from long-standing relationships between markets are emerging, possibly suggesting that a particular market is overbought or oversold. There may also be subtle interactions between markets whereby certain markets may lead other markets [40]. Globalisation of companies and capital markets increases the links between the performance of individual equity markets, with many smaller equity markets taking general direction from major markets such as the US. Examples of intermarket indicators which can be relevant in predicting the value of an equity market index include bond prices, stock indices in other countries, and commodity prices such as oil.

When incorporating intermarket data into a trading model, it is important to ensure that future data is not accidentally supplied to the model. Consequently, care must be taken when using data drawn from markets in different time zones. Although the date attribute of discrete pieces of input data might appear to agree, a model may actually be including future information from a later time zone in making its predictions, thus biasing the performance of the developed trading system.

Fundamental Indicators

The choice of relevant fundamental indicators will depend on the financial market which is to be traded. Fundamental analysis can be applied to individual firms in an attempt to assess whether their share prices are currently under or over-valued, or it can be applied at a macro-economic level to assess the likely performance of, for example, the equity market in its entirety.[1]

If the intention is to assess individual firms, useful fundamental indicators will include current and historic information on dividends, profits, sales, assets, debt levels, and liquidity. These factors, along with non-financial information

---

[1]An interesting parallel can be drawn here with horse racing. Some gamblers analyse fundamental factors such as a horse's past racing history and information concerning jockeys, trainers, ground conditions in order to determine whether the odds offered on a particular horse in a race are mis-priced. Such mispricings can give rise to good-value betting opportunities. Other gamblers concentrate their attention on what they observe in the betting markets (technical analysis). A good discussion of both these groupings of strategies is provided in [69].

about the firm, could be compared with similar information on the firm's competitors to assess the firm's competitive position versus its peers. This information can be combined with data from a sectoral and macro-economic analysis to form an assessment of the likely future growth potential for the firm's profits and dividends.

If fundamental analysis is being performed for the equity market as a whole, the primary indicators will include the broad drivers of supply and demand in the economy. These impact on the earnings and dividend potential of firms, therefore impacting on financial asset prices. Examples of these indicators include:

- commodity prices,
- term structure of interest rates,
- foreign exchange rates,
- GNP,
- inflation,
- rate of unemployment, and
- budget/trade deficits.

These factors have a non-linear and time-lagged impact on each other, on the value of individual shares, and on the equity market as a whole. In the latter case, the impact of a change in a fundamental indicator on a market index will vary over time as the firms comprising the market index change.

Several practical problems can arise when using fundamental indicators drawn from macro-economic data. Fundamental indicators by their nature are time-lagged. Hence, the unemployment rate for March will not be known until figures are compiled in April or May, depending on the speed of collection of the data. In using information drawn from historical databases, the modeller must ensure that data is only presented as input to a model when it was actually available; for example if unemployment data was an input to a model, the input for April would actually be the lagged rate of unemployment in (perhaps) March. If government statistics are being included as fundamental indicators in a model, their definition and measurement should be consistent over the period of interest. This is not always the case. Another problem that can arise when using such data is how to deal with subsequent revisions of the data. This can lead to a problem when using historical databases which consist of clean (post-revision) information. A model constructed using clean data may prove brittle when exposed to poorer-quality real-world data.

EC Approaches to Using Fundamental Indicators in a Trading System

In the earlier discussion of the GA (Chap. 3), an illustration was provided of how screening rules to determine which financial assets should be purchased (or sold) could be evolved from collections of fundamental or other indicators.

An alternative way of constructing a stock-picking rule is to encode 'levels' of a group of indicators on a string. For example, suppose the intent is to encode different combinations of fundamental indicators on a binary string. The first bit could encode whether the debt level of a firm was high or low relative to the industry average, the second bit could indicate whether the firm had experienced above-average industry sales growth over the past three years and so on for other fundamental indicators.

High sales growth relative to industry average?	High debt level relative to industry average?	High level of cash flow from operations relative to industry average?	High level of liquidity relative to industry average?	High profit level relative to industry average?

Fig. 8.2. String encoding of a number of fundamental indicators. Each indicator can be coded as a 0 (no) or 1 (yes)

If we restrict attention to a case where the rule consists of, for example, 25 binary decision variables, the total number of possible rule combinations is 2^{25} or 33,554,432. It is clearly difficult to exhaustively examine all of these, hence an evolutionary algorithm like the GA can be used to determine a good stock screening or trading rule which combines these fundamental indicators. A population of binary strings each representing a specific trading rule could be created randomly, with the GA then selecting which of these represent a good *screening rule* for investment purposes, and then applying crossover and mutation to uncover yet better trading rules. The fitness of each rule could be tested on historical data: if rule x had been applied over the past y time periods, what risk-adjusted performance would it have generated?

8.2.2 How Much Data Is Enough?

There is no simple answer to this question. Generally, the more relevant data the better. Use of small datasets increases the risk of model overfit on the training data, with poor generalisation out-of-sample. The issue of how much data is required is bound up with the number of parameters being estimated in the model. An old saw in statistics is that a modeller should have at least 10 data observations for each included parameter. Hence, if a complex model is being constructed, for example a MLP with many weights, a considerable quantity of data may be required to robustly train the model. The *curse of dimensionality* points out that the amount of data required to construct a model increases exponentially with the number of parameters in the model. To gain intuition on this point, consider what happens if a modeller increases the number of explanatory variables in a linear regression model for a fixed-size dataset. As the dimensionality of a model increases, the coverage of the data space by the fixed-size dataset is reduced. The data points separate further from each other in the expanded data space.

In financial applications, the quantity of data available will vary depending on whether the model is being constructed with daily data or with intra-day data. If daily data from a stock market is being used, approximately 250 values will be available each year. If intra-day tick-by-tick data is being used the number of data values available for a calendar year will be considerably greater. A problem with using daily or lower-frequency data is that if data is drawn from a long time period, say 15 years, it is questionable as to whether the market has remained unchanged over that period. The underlying data-generating process for market data, unlike that for physical processes, is not stationary. If a modeller believes that only recent data is likely to be useful, perhaps because of significant recent changes in market regulation, but is concerned that this leaves a small dataset, one way to overcome the problem is to create a larger *synthetic* data series from the data available. On the assumption that small changes in the inputs should produce relatively small changes in the value of outputs, new (synthetic) data can be created by taking existing input-output data vectors, making small random changes to the inputs, keeping the same output value as the original data vector, and adding these new data vectors to the original dataset.

The intended lifespan of the model before it is retrained will also have an impact on the quantity and form of input data required. If a model is only to be used for a short period before it is replaced, it will not need to be able to detect long-term market trends, therefore simplifying the data preprocessing requirements. However, the risk of such a model is that it will be fragile with respect to changing market conditions.

Penalising Model Complexity

In considering issues of data sufficiency and the number of explanatory variables to include, the lesson of *Occam's razor*[2] should be borne in mind. If there are two competing explanations for an event, all other things being equal, the simpler one is to be preferred. In order to control model size when building a model, the error criterion can be adapted to incorporate a penalty term based on the model's complexity. This term acts to reduce or penalise the model's performance as the fitted model becomes more complex. A wide variety of metrics from the traditional statistical literature on model selection can be employed to manage the trade-off between model fit and model complexity, including Akaike's Information Criterion (AIC), minimum description length (MDL), and Schwarz's Bayesian Criterion (SBC).

---

[2]The philosopher William of Occam (approx. 1280-1347) is reputed to have said 'Entia non sunt multiplicanda praeter necessitatem' ('Entities should not be multiplied more than necessary').

8.3 Selecting and Preprocessing the Data

Once a plausible set of explanatory inputs has been selected based on financial theory and intuition, the task facing a modeller is to determine which of these inputs should be incorporated into the final model, and how the included inputs should be preprocessed to extract the maximal useful information content. The steps of input selection and preprocessing are intimately interlinked in practice but are separated below for ease of discussion.

8.3.1 Selection

In selecting the inputs, the first step is to perform a *data audit* to identify missing data, and to filter incorrect data before data analysis starts. Methods for this include graphically examining the data series, checking for logical inconsistencies in the dataset (including cases where the closing price for a period is higher than the high-price for the same period, or where the opening price for a period is less than the low-price for that period, or where a zero price is recorded), and the calculation of simple descriptive statistics for each series (mean, maximum, minimum, standard deviation, and the number of data items). These steps will help identify possible outliers in the dataset, and will also help build the intuition that the modeller has concerning the raw data that is being analysed.

Once the data has been cleaned, traditional statistical techniques can help in deciding which data series are suitable candidates for inclusion in the model. Basic tools such as examining the correlation of potential inputs with each other and with the target output can help rule out inputs with little information content, for example data which is invariant with respect to the predictive target, thereby focusing attention on plausible useful inputs. Linear regression models can be constructed between sets of inputs and the target output to identify useful inputs (do the regressions yield coefficients which are significantly different from zero?). Factor analysis techniques can be applied to reduce the number of inputs supplied to the model, by compressing multiple inputs into a limited number of principal components. A drawback of these techniques is that they will not uncover non-linear relationships between inputs and outputs, and hence they can only provide suggestive rather than definitive guides as to which inputs should be eliminated as having little apparent information content.

It may also be useful to create new, additional data series for possible inclusion in the predictive model. For example if the financial market of interest is known to be seasonal (as are many commodity markets), it is often useful to add a seasonality variable as an input.

8.3.2 Preprocessing

One of the most time-consuming aspects of financial modelling is the preprocessing of model inputs in order to make patterns in the data easier for the predictive engine to detect. Preprocessing has two main stages, the transformation of inputs and their normalisation.

Transformations

Data transformation can be described as the redefining of data using a predefined rule. Common reasons for transforming data include the compression of multiple raw inputs, in order to produce a single model input, and the removal of some aspect of the data (for example a long-term trend) in order to concentrate attention on another characteristic of the data instead.

Transformations to compress the data allow a reduction in the number of inputs presented to the model. A simple example of such a transformation is to use the *ratio* of two pieces of data (rather than the raw data). For example the ratio of the number of advancing versus the number of declining shares on the stock market can provide a more useful indicator of market sentiment than would either measure on its own. Other common transforms are to use *differences* such as today's value less the value of x periods ago, the percentage change in price between two dates, or moving averages which act to compress time-series information in order to smooth out noise and uncover longer-term trends in a data series.

An example of a transformation which concentrates attention on one aspect of the data rather than another, is the removal of a long-term trend in a time-series. If the intention is to forecast over a short-run horizon, it is often useful to transform the time series of interest by subtracting a y-period moving averages of their values from its current values (Fig. 8.3). If a long-term moving average is subtracted from the current price of a financial asset, it eliminates the longer-term trend and emphasises shorter-term swings in the data.

More complex transformations of raw data in a time-series of price information can be undertaken, such as the calculation of the rate of change of a moving average or the use of first-order log difference of the price changes. The rate of change of a moving average can be obtained from the regression:

$$Y_{t-x} = \alpha + \beta P_t \tag{8.4}$$

where P_t is today's price and Y_{t-x} is the moving average over the last x days. The value of β represents the sensitivity of the moving average to a change in today's price. The first-order log difference of the price changes is given by:

$$O_i = \ln\left(\frac{P_t}{P_{t-x}}\right) \tag{8.5}$$

where P_t is today's price, P_{t-x} is the price x days ago, and O_i is the first-order log difference. Technical indicators provide other examples of input data transformations.

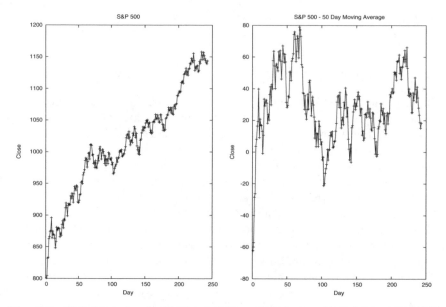

Fig. 8.3. S&P 500 index for 11 March 2003 to 25 February 2004 (left) vs S&P 500 index less a 50-day moving average of the index for the same period (right)

Transformations can also be applied to the predictive target and this step will be required if the range of output from the modelling methodology is constrained. For example if an MLP is being used, the form of transfer function at the output node will determine the numerical range of output which the network can generate. If a logistic or tanh transfer function is used, the outputs are limited to the ranges (0,1) and (-1, +1) respectively. However, neither function is sensitive at the extreme lower or upper limits of its output range, hence it is usually appropriate to rescale target outputs into a narrower range such as (0.1, 0.9) for the logistic function, or (-0.9, +0.9) for the tanh function (Fig. 8.4).

Normalisation

Normalisation is carried out to distribute the raw input data more evenly across its range of variation, and, if necessary, to scale it into an acceptable range for the modelling technique being utilised. An initial step may be to examine the data in order to determine which ranges of it the modeller wishes to focus attention on. If a data series has a small number of outlier (extreme) observations it may be appropriate to clip the values at +/- 2 standard deviations above and below the mean of the series. The object of clipping the data is to focus attention on the range of typical values that the input assumes most of the time, thereby making the model more sensitive to changes in the inputs in their usual range. If, after the outliers have been removed, the remaining

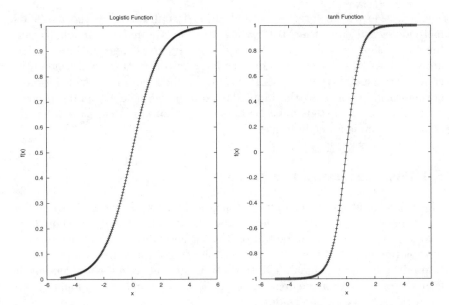

Fig. 8.4. Logistic and tanh functions

data is concentrated across a narrow range of values it is usually useful to rescale it so as to spread out the data to allow the modelling technique to more easily detect differences in the input. The object of this step is to make the distribution of the input data series more uniform. Simple mathematical tricks which can help in this task (depending on the nature of the raw data) include applying the log or exponential function, squaring the data, or taking the square root of the data.

Rescaling each input into a standard range can help ensure that each input has an equal chance to influence the output of the model. The initially selected inputs for the model may have widely differing magnitudes; the current value of the S&P 500 may be 1250, whereas the US\$-Euro exchange rate may be 1.2. By rescaling all the inputs into a fixed range, (0,1) or (-1, +1), the model can more easily place appropriate weight on each. A simple method of carrying out this step is to apply linear scaling. For example, to rescale a value at time t into the range $(0 \rightarrow 1)$, using data from the last n periods:

$$x_{rescale} = \left(\frac{value_t - lowest_n}{highest_n - lowest_n} \right) \tag{8.6}$$

To rescale into the range (-1 +1), the above formula is adapted by subtracting 0.5 and multiplying the result by 2. An alternative method of normalisation is to rescale based on the mean and standard deviation of the data over the last n periods:

$$x_{rescale} = \left(\frac{value_t - mean_n}{\sigma_n} \right) \tag{8.7}$$

The selection of the form of normalisation is influenced by the modelling methodology. Taking the MLP as an example, the most common transfer functions (sigmoid and tanh) are most sensitive to changes in input values around -1 to +1. Hence, rescaling of inputs into this range will tend to increase the sensitivity of the MLP. Whatever the form of transformation or normalisation applied to the input data during the model training process, obviously the same steps must be applied to preprocess live data when the trading system is actually in use.

8.4 Postprocessing the Output

The ultimate output of a trading system is a signal corresponding to the action the investor should take. Should a stock be bought or sold, or should the investor stay out of the market altogether? Therefore it is necessary to postprocess the output of the predictive engine component of the trading system to produce a trading signal. The trading signal generated for a given output from the predictive engine depends on the *entry, exit* and *money management* strategies selected by the trader.

8.4.1 Entry Strategy

An entry strategy determines what output is required from the predictive engine before the system generates a buy (or short sell) signal, enters the market and takes a position. An example of a simple entry strategy would be to buy a share once a trading system predicts it will rise by 3% or more in the next ten days. More sophisticated entry strategies could vary the amount invested depending on the strength of the trading signal produced. Entry filters could be built into the trading system to ensure that trades are in round-lots and of a minimum size, to avoid small cost-inefficient purchases being made.

8.4.2 Exit Strategy

The exit strategy determines when trading positions should be closed out. A position may be closed out in order to capture gains on a trade or to protect against excessive losses when a trade goes wrong, or a trade may be exited because the market has turned. An investor can reduce trading risks by using *stop-loss* and *take-profit* triggers. Under a stop-loss trigger, a position is sold out once a loss of $x\%$ or of $\$x$ occurs.

The selection of a suitable stop-loss trigger can be undertaken judgementally[3] or by simulation. In the latter case, the trader could simulate the performance of the trading system on training data with no stop-loss trigger

---

[3] As an example of how judgement could be applied, Osler [179] provides evidence for the clustering of stop-loss and take-profit orders in foreign-exchange markets. The

included, and produce a graph of maximum drawdown for each trade versus the final profit outcome on that trade. This will allow estimation of the frequency that a drawdown of x or more was followed by a final trading profit, and this can provide an input into the selection of a stop-loss trigger for the system. Under a take-profit trigger, a gain is realised once a profit of $y\%$ or y occurs.

Many variants on the simple take-profit trigger exist, including a *trailing stop* where a position is exited once a given level of profit has been achieved, and a price fall then occurs (the market reverses). Alternatively, an exit can be triggered when a subsequent sell signal is indicated by the trading system, or after a set period of time if the take-profit trigger has not been hit.

Although the use of stop-losses can protect a trader, they will sometimes fail. Consider the case where a company announces bad news just before the market opens. The effect of this could be to cause the share price to *gap* substantially downward at the market open, causing the price to fall below the stop-loss trigger before it can be activated (Fig. 15.3 for an illustration of an intra-day gap). A stop-loss is a risk-management, not a risk-insurance tool. Other risk-management tools include the use of put and call options.

8.4.3 Money Management

Money management strategies include limiting the amount of money risked on a single trade, a single stock/sector, or indeed on a single trading model. Investors do not have access to infinite funds, so the drawdown characteristics of trading systems is important. Generally, successful trading strategies should produce good returns, a smooth increase in the *equity curve* (the cumulative profits generated by the trading system over a time period), and little clustering of losses (limited drawdowns) [87]. An example of an equity curve is provided in Fig. 8.5.

8.5 Validating the System

Once a prototype trading system has been developed using training data, it must be rigourously validated before going live. Trading models are typically *back-tested*. They are constructed using historical market data and the

---

affect of clustering of these orders is that trends in exchange rates tend to behave predictably at these cluster levels. If a stop-loss level is hit, the downward trend in price will tend to accelerate as many stop-losses are hit simultaneously. Similarly, if an upward trending price hits a cluster of take-profits, the upward trend will tend to halt as many investors take profits at this point. Testing this hypothesis, Osler found clusters of take-profit orders at exchange rates characterised by 'round numbers' (ending in 00) with stop-loss orders clustering just beyond round numbers. It is suggested that equity-market stops should be set just above round-number prices [119].

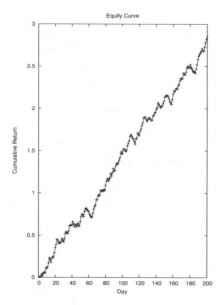

Fig. 8.5. An example of a smooth equity curve from a trading system. Returns are scaled in $000

risk/return characteristics of the trading system are examined to determine whether they are in line with the trader's preferences. Once the model has been constructed, it is further tested using historical data which was not used to train the model. The purpose of the out-of-sample testing is to determine how robust the model is. Do the training period results generalise well to the out-of-sample period? When splitting data into training and out-of-sample datasets, several methods can be adopted. The simplest is a fixed split, for example 60% of available data is used to train the model and the remaining 40% is used to validate it.

An alternative to back-testing which can be used to validate a trading system, particularly one which trades on a short-term horizon, is to perform shadow walk-forward testing. The system is provided with a real-time data feed, and executes simulated trades based on this live information. The performance of the system in shadowing the current market is then assessed.

Limitations of Back-testing

Although back-testing is an essential validation tool, the limitations of the technique must be remembered. Any trading model constructed and tested using historic data will tend to perform less well in a live environment than in a back-test period for a number of reasons. Live markets have attendant problems of delay in executing trades, illiquidity, interrupted/corrupted data and interrupted markets. The impact of these issues is to raise trading costs and consequently to reduce trading profitability. In addition, markets are competitive and represent a *Red Queen*: as one market participant introduces new computational technologies in an attempt to gain a trading edge, other traders rapidly imitate the technique to erode its profit potential [88]. Hence, estimates of trading performance based on historical data may not be replicated in live trading as other market participants will apply similar technologies. Also, if the particular trading strategy had actually been implemented in the back-testing period, the trading activity and consequently the prices for securities could have been affected, reducing the actual profitability of the system.

Examining the Test Results

Whether the system is tested using historical or current market data, the characteristics and results of its trades must be carefully examined. The goals for a trading system are usually a balance of generating good returns at acceptable risk. Once the system is constructed, its performance and the robustness of its performance in both the training and out-of-sample periods can be evaluated across several metrics. Measures of return which could be used to evaluate the model's performance include the:

- total profit over a specified period,
- number of trades,
- win-ratio (percentage of profitable trades),
- average profit per trade,
- average profit per successful trade,
- average loss per losing trade,
- the profit factor $\left(\frac{\text{total profit on winning trades}}{\text{total loss on losing trades}} \right)$.

Measures of the robustness of the trading system include:

- standard deviation of return per trade,
- Sharpe ratio,
- modified Stirling ratio,
- maximum drawdown, and
- maximum profit (loss) on a single trade.

Although risk-adjusted trading profits will be a prime metric, the modeller will be keen to see how these profits are generated. A visual examination of the equity curve will help reveal when profits and losses are generated by the trading system (under what market conditions do each occur?). For example, Fig. 8.6 suggests a case where the trading system worked well for the first 100 days trading, but its performance deteriorated notably after that suggesting that the system needs retraining.

The modeller will also need to determine whether the profits are generated over many trades or are generated from a very small number of big wins?

What about the losses? Any big losses? A histogram of the distribution of trade profits and losses will highlight any unusual distribution of these items. If large profits or losses occurred, can the modeller determine why they might have occurred? For example, was company-specific news released that day?

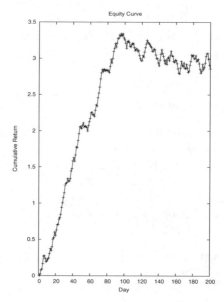

Fig. 8.6. An equity curve which suggests that the model needs retraining. Return axis is scaled in $000

In looking at the profits (losses) from individual trades, some rough statistics can be calculated to help assess whether the trading system produces results which are significantly better than chance alone. For a given sequence of trading results, a t-test can be performed, to determine whether the average profit per trade (assuming there is one!) is significantly better than zero. However, simple statistics should be used with caution. There is likely to be serial dependence in the sequence of trading results as all the data is drawn from a single time period, not randomly, which reduces the effective sample size. Another problem is that if a number of trading systems are created, and the developer is trying to choose the best trading system amongst them, it is likely that one or more trading systems will exhibit results which produce a high t-value through chance alone, and multiple-comparison statistical methods should be used.

Another issue in validating the model will be to test whether the exit strategies for the system could be altered to filter out any major losses which have occurred? Did the 'take-profit' trigger leave substantial profits on the table on several occasions? If so, should the trigger point be raised?

The characteristics of the trading system must be similar on both the training and out-of-sample datasets before they are considered to be robust. Once the system goes live, the above diagnostic metrics should be captured on a regular basis and compared with the results from both the training period and earlier live trading periods, to highlight any degradation in the system's characteristics. Shifts in these metrics can provide an indication that the system should be replaced.

When powerful modelling technologies are applied to a dataset, there is always the danger that they could uncover a spurious pattern in the data which just happens to work well for a time. Apart from using extensive out-of-sample testing to reduce this possibility, the system should pass an *intuition test*. Although not all biologically inspired methodologies, for example MLPs, are amenable to easy deconstruction, rules generated by GA or GE models can be examined. If the rules make no sense to a domain expert, this suggests that they are likely to be spurious. The sensitivity of the model to small alterations in its parameters (for example, its lag periods) should also be checked. If small changes in these parameters result in large changes in the model's performance, the model's results should be viewed with suspicion.

In the above discussion of model validation there has been an assumption that a single (static) trading model has been constructed and then traded for a period of time. An alternative (and common) methodology is to use a *moving window* approach, where a model only predicts one step (or a small number of steps ahead) at a time (Fig. 8.7). The model is continually retrained as new data becomes available. The use of a moving window training implies that the trading model being used changes or adapts over time (see Chap. 14 for an example of this approach).

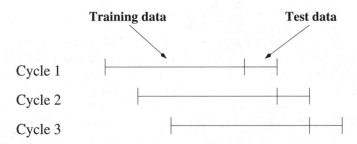

Fig. 8.7. A moving window approach. The system is trained and then tested. After a period of N days (here the length of the test data period), the system is retrained and the training data window is advanced by N days

Iterating the Model Development Steps

The validation and earlier model development steps will be undertaken iteratively. It will usually be possible to improve the performance of the first trading
model by careful diagnosis of its errors. Are all the inputs actually influencing
the model's decision? Do the results degrade substantially if any of the inputs
are dropped? Can the results be improved by altering the predictive target,
the input data, or the data pre-/postprocessing steps? A considerable time
may be spent iterating between model validation and model development.

8.6 Implementation and Maintenance

Once the model has gone live, a further decision faces the modeller. How long
should the model be used before being scrapped? There is no easy answer to
this question. Obviously, if a major market event occurs, for example a war, it
is quite possible that this will render the assumptions underpinning existing
trading systems redundant. The development of new financial products can
also produce a change in market structure. For example, significant changes
were noted when listed options trading began in the US in 1973, and when
stock index futures trading began in 1982. A change in the regulatory environment governing the market of interest can also have implications for the
performance of trading systems. Examples of regulatory changes include the
introduction of *circuit breakers* on the NYSE, and the move to full decimal
pricing on the NYSE on 29 Jan 2001. A short discussion of each of these events
is provided to illustrate how rule changes can impact on trading systems.

Circuit Breakers

In response to the market crashes in October 1987 and October 1989, the
NYSE instituted several circuit breakers to reduce market volatility. Under
Rule 80A, when the Dow Jones Industrial Average (DJIA) moves 180 points
or more (the size of this collar is defined each quarter, based on 2% of the
average closing of the DJIA for the last month of the previous quarter) from
the previous day's close, index arbitrage orders in component stocks of the
S&P 500 stock index are subject to a 'tick' test. In up-markets, buy orders
may only be executed on a minus or zero-minus tick, in down-markets sell
orders may be executed only on a plus or zero-plus tick. Once activated,
the rule applies for the rest of the day, unless the DJIA moves back within 90
points of the previous day's close. Rule 80A was activated 51 times on 47 days
in 2001. It has been widely credited for helping to reduce market volatility.
Under Rule 80B, if the market falls by 10, 20 or 30%, (based on the average
closing of the DJIA for the last month of the previous quarter), a market-
wide trading halt is activated. The length of the halt varies from 1 hour for a
10% decline to the remainder of the day for a 30% decline. Trading systems

which took no account of market regulations when they were back-tested, or which were developed before significant rule changes occurred, could perform unexpectedly when faced with real-time effects of these rules.

Decimal Pricing

Historically, shares on the NYSE were priced in eighths (of a dollar), and in sixteenths since 1997 [9]. The move to full decimal pricing was completed in 2001. Studies undertaken after this pricing change occurred indicated that it had resulted in a tightening of the bid-ask spread (the difference between the buyer's bidding price and the seller's asking price), from a trade-weighted average of 17c per share in 2000 to approximately 8c in 2001. This change in market structure could clearly impact on the profitability of a trading system constructed before the pricing rule alteration.

Monitoring the System

Even leaving obvious shocks aside, market conditions and the utility of any trading system will change over time. All trading systems embed a simplified representation of the real-data generation process of financial markets, and omit many relevant variables. The significance of the omitted variables will change over time. One rule of thumb is that a short-term trading system should be retrained when 10-15% of the training data can be replaced with new data. Another approach is to construct a monitoring system which looks for degradation in the trading system's performance. A simple monitoring system would track recent trading performance and compare it with the performance in earlier time periods. A change in the trading characteristics of the system, even if the returns from the system are still satisfactory, may provide an early-warning that current market conditions are diverging from those on which the model was trained.

Diversification

A tried-and-tested strategy to deal with market risk is to diversify investment funds across trading systems and across markets. No single trading system can be expected to work well under all market conditions. To overcome this problem multiple systems could be constructed, each of which impounds a different representation of the market. A multi-stage trading system could be constructed, where the first stage (a *gating mechanism*) classifies the 'type' of market that exists at present, and then uses this to select which of a portfolio of trading systems to use. For example, moving average techniques work poorly in non-trending or *choppy* markets. Hence the system could turn off moving average-based components of the trading system if such market conditions are detected. Alternatively, rather than using a hard gating mechanism where a model is either used in full (100%) or not at all (0%), a soft gating mechanism could be applied which weights the output of each individual model depending on the current market type, and combines their individual outputs to create a 'community' trading decision.

Of course, there is no requirement that a trading system should seek to invest in all conditions. Under the *coherent market hypothesis* [216] markets may be more predictable in certain phases of the business cycle than others, and sometimes it may be better to abandon trading efforts and await more benign conditions.

8.7 Summary

The selection and implementation of a specific biologically inspired algorithm is only one component in the process of developing a complete trading system. No algorithm can compensate for poor-quality data or a poor trading system design. Earlier in the chapter, one of the major sources of information for short-term trading systems (technical indicators) was introduced. The next chapter discusses these in more detail.

9

Technical Analysis

Technical analysis can be defined as the attempt to identify regularities in the time-series of price and volume information from a financial market [144]. Technical analysts contend that through study of the relative strength of the current forces of demand and supply in the market, and the identification of zones of price support and resistance, they can gain insight into the likely future trading range and direction of price movement for a financial asset. Technical analysts also believe that financial asset prices move in trends, and that price patterns repeat themselves [160]. Although this chapter will discuss the methods of technical analysis in the context of equity markets, the concepts can be applied to any traded financial market.

Technical Analysis and the Efficient Market Hypothesis

The position of technical analysts on market efficiency is often incorrectly stated. Technical analysts do not consider that the market is inefficient in processing available information into prices, rather they agree that prediction of market prices is difficult. Strict technical analysts argue that fundamental analysis is unlikely to generate risk-adjusted excess-returns, as markets react quickly to new information. Due to their pessimism that fundamental analysis can uncover new information which is not already impounded in asset prices, technical analysts choose to study the effects rather than the causes of market price movements. In essence, technical analysts attempt to anticipate what other market players are likely to do next. Based on recent price/volume information, which stocks are likely to be in or out of favour with investors in the near future?

Although considerable controversy exists amongst financial theorists regarding the veracity of the claims of technical analysts, their methods are widely applied in practice, and the language of technical analysis permeates financial newspapers. In a study conducted by Taylor and Allen [211] on behalf of the Bank of England, it was found that approximately 90% of financial institutions dealing in foreign exchange in London placed some weight on information obtained from technical analysis. Greatest use was made of this information in forming short-run exchange rate expectations. Research in a variety of financial markets has provided at least tentative support for the utility of technical analysis. In a study of technical analysis in stock markets, Brock, Lakonishok and LeBaron [35] found suggestive evidence that simple technical trading rules had predictive power and concluded that the findings of earlier studies that technical trading rules did not have such power were 'premature'. In the case of foreign-exchange markets, studies which suggest that technical trading strategies may be profitable include Sweeney [210] and Levich and Thomas [139].

9.1 Technical Indicators

The development of trading rules based on current and historic market price (or exchange rate) and volume information has a long history [36]. The process entails the selection of a set of technical indicators and the development of a trading system based on these indicators.

The core concept in technical analysis is that of a *trend*, with future prices considered to be more likely to continue to move in the direction of the current pervasive trend than to reverse. The objective of trading systems developed using technical indicators is to identify the current pervasive trend and to trade in that direction, until a trend reversal is anticipated [160]. Technical analysts argue that trends indicate a relative imbalance of supply and demand for the financial asset, and that they are more likely to continue than to reverse, giving rise to market lore such as *the trend is your friend* and *never buck the trend*. Of course, to the extent that investors are disposed to engage in herd behaviour as suggested by behavioural finance,[1] we would expect market trends to result.

In assessing the current trend in the market, technical analysts consider that there is no single trend active in a market at a point in time, but that there are multiple trends of differing periodicity (long, medium and short run),

---

[1]Lintner (1998) [142] defines behavioural finance as being 'the study of how humans interpret and act on information to make informed investment decisions' (p. 7). The field merges concepts from financial economics and cognitive psychology in an attempt to construct a more detailed model of human behaviour in financial markets. Adherents to the behaviouralist school believe that decision-makers (investors) do not always behave in a strictly rational fashion, and that their departures from rationality stem from the nature of the mental models that people use to analyse complex environments. The resulting biases in decision-making can produce regularities in investor behaviour including over (under) reaction to price changes or news, extrapolation of past trends into the future, lack of attention to the fundamentals underlying a stock's valuation, undue focus on popular stocks, and seasonal price cycles.

which may be moving in different directions, combining together to produce currently observed price movements. Agreement between trends at different levels of periodicity is considered to indicate a stronger trading signal than when the various trends do not agree.

Identifying Trend Reversals

If the objective is to trend-follow, identification of possible trend reversal price-points is clearly important. Concepts which are relevant to this task include those of *support* and *resistance*. A zone of support (or a *floor*) is a price level that a financial asset has reached but not fallen below. Resistance (or *the wall*) is a price level that has been reached but not breached. Therefore a zone of support arises at prices where there is a concentration of demand, a zone of resistance arises when there is a concentration of supply. The rationale behind these ideas is that support levels represent the minimum price that a large number of current holders of the stock are willing to accept, whereas resistance represents the maximum amount that a sizeable body of potential buyers are willing to pay. Technical analysts suggest that down-trends in a price series tend to reverse at zones of support, whereas up-trends tend to reverse at resistance zones. Therefore, stocks should be sold when their price bumps into their resistance level, and bought if their price bounces off their support level.

If these zones are breached a *breakout* occurs, signaling that something unusual in terms of the stock's recent price history has occurred, perhaps because of a significant new information flow to the market. Once the breakout occurs, technical analysts consider that the trend in price movement will accelerate and new support and resistance levels will be established. Therefore, shares breaching a resistance level should be bought, with those falling through a support level being sold. Figure 9.1 provides an illustration of the concepts of support, resistance and a breakout.

Behavioural finance explanations can be offered for the concepts of support and resistance. If a share is trading in a defined range for a while, and moves lower towards the support level, some investors may perceive that it is now good value and buy. Conversely, if it moves back towards a resistance level, investors who previously bought at that level may wish to sell out as their expectations of making a profit on their purchase were unfounded [156, 197, 213].

Categories of Technical Indicators

Although there are potentially an infinite number of technical indicators which can be formed from historic price and volume information, the financial literature [35, 160, 182] suggests that four groupings of indicators are widely used to create *entry signals* by investors (the reversal of a previous entry signal generated by one of these indicators may also trigger an *exit signal* for investors).

Fig. 9.1. Example of support, resistance and breakout, using IBM intra-day data

In essence, all these methods perform an empirical time-series analysis based on price and/or volume data. The four groupings of indicators are:

- moving average indicators,
- momentum indicators,
- breakout indicators, and
- oscillators.

9.1.1 Moving Average

The simplest moving average filter systems compare the current price of a share, or an exchange rate, with a moving average of the price or exchange rate over a lagged period, in order to determine how far the current price has moved from an underlying trend. As they smooth out daily price fluctuations, moving averages can heighten the visibility of an underlying trend.

Hence a moving average serves as a *low-pass* filter, in that it transmits low-frequency signals (trends), but dampens out high-frequency price fluctuations. Figure 9.2 graphs the daily closing price for the Dow Jones Industrial Average against its moving average for 10 and 200 days. It is noticeable that the longer moving average is much less volatile than either the shorter moving average or the actual daily value for the index.

Moving average indicators can be used either to construct a trend-following, or a counter-trend trading system which aims to anticipate trend reversals. Here we will concentrate on the trend-following variant. A buy trading signal to close out short positions and/or go long can be generated when the current price moves above its moving average, and a sell trading signal (close out long positions and/or go short) if the current price goes below the moving average. The intuition behind this idea is that if the price has moved above its moving

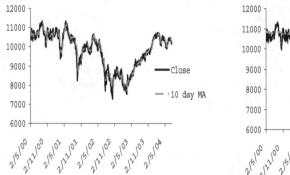 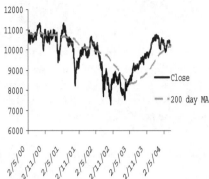

Fig. 9.2. DJIA for 2/5/00 to 2/5/04, with 10-day (left) and 200-day moving average (right) superimposed on the closing price

average, this suggests that buying pressure has emerged and that a bullish trend in the price is occurring. Variations on this approach include the use of an exponentially smoothed moving average which gives greater weight to more recent data.

MACD Oscillator

Another common use of moving averages is to construct a moving average convergence-divergence (MACD) oscillator, calculated by taking the difference of a short-run and a long-run moving average. If the difference is positive, it is taken as a signal that the market is trending upward. For example a buy signal could be generated when the shorter moving average crosses the longer moving average in an upwards direction. A sell signal could be generated in a reverse case. Therefore, a sample MACD trading rule could be:

IF x-day MA of price $\geq y$-day MA of price

THEN Go Long ELSE Go Short

where $x < y$ (for example $x = 10$ and $y = 50$). The MACD oscillator is a crude band-pass filter, removing both high-frequency price movements and certain low-frequency price movements, depending on the precise moving average lags used. In essence, the choice of the two lags produces a filter which is sensitive to particular price-change frequencies.

In a recursive fashion, more complex combinations of moving averages of values calculated from a MACD oscillator can themselves be used to generate trading rules. A sample trading rule in this case could be:

IF x-day MA of price $\geq y$-day MA of price AND

$$a\text{-day MA of price} \geq b\text{-day MA of price}$$

THEN Go Long ELSE Go Short

Even if such an input to a trading model is useful, there is no guarantee that the optimal length of the two moving averages will be constant over time. A more complex approach is to create several MACD oscillators and let the trading system vote, based on the output of each oscillator whether to generate a buy or sell trading signal.

Drawbacks of Moving Average Indicators

As moving average indicators are trend-following devices, they perform best in trending markets. They can have a slow response to changes in trends, missing the beginning and end of each move. They also tend to be unstable in sideways-moving markets, generating repeated buy and sell signals (*whipsaw*) leading to unprofitable trading. Trading systems using moving averages tradeoff volatility against sensitivity. The objective is to select the lag period which is sensitive enough to generate a useful early trading signal but which is insensitive to random noise. One strategy is to use an *adaptive moving average* whereby the period of lag adjusts depending on the level of volatility in the market. In non-trending markets, a slow (long period) moving average is used in order to avoid the whipsaw effect.

Another difficulty which can emerge in using moving average indicators is that the longer the period of the moving average used, the greater the quantity of data required for model building and testing.

9.1.2 Momentum

The momentum, or rate of change, of an asset price is the ratio of a time-lagged price to the current price:

$$\frac{Price_t}{Price_{t-x}} \tag{9.1}$$

Momentum can also be expressed as a percentage:

$$\frac{Price_t - Price_{t-x}}{Price_{t-x}} \tag{9.2}$$

The belief underlying this indicator is that a strongly tending price is likely to persist for a period of time as more investors seek to buy or sell the trending share. There is recent evidence that momentum trading strategies can work, particularly when investing in smaller firms [109]. It is also argued by some analysts that *momentum precedes price* in that price momentum peaks before the absolute price peaks [186]. A reduction in upward momentum indicates that an upward trend is weakening and the market is becoming overbought, and may provide a leading signal that a trader should sell. Conversely, weakening negative momentum indicates that the market is becoming oversold, and provides a buy signal.

9.1.3 Breakout

There are several forms of breakout trading models. The simplest are trend-line breakouts, usually displayed graphically. If the price of a financial asset breaks through a trend-line (a previous resistance level) from below, a long position is taken. If prices drop below a previous support level, a short position is taken.

Channel Breakout

A related approach is a *channel breakout model*, which combines the idea of both support and resistance. When the price of a financial asset moves out of its recent trading range (defined by its maximum or minimum value in a lagged time period), an entry signal is indicated. A simple example of a trading rule derived from a breakout indicator would be to buy a financial asset when it exceeds its previous high in the last four weeks, and conversely to sell short (or if holding the asset already, sell it) if it falls below its previous four week low. A simple trading rule based on this idea is:

$$IF\ C(t) \geq\ Maximum[H(t, \ldots, t - 20)]\ THEN\ Go\ Long$$
$$ELSE\ IF\ C(t) <\ Minimum[L(t, \ldots, t - 20)]\ THEN\ Go\ Short$$

where $C(t)$ is the closing price of the asset on day t; $H(t)$ and $L(t)$ are the highest and lowest prices respectively, of the security in the last 20 trading days.

If the market is trending, the trading range will also change over time. A more complex approach is to plot an envelope above and below the most recent price. One well-known version of this is to calculate *Bollinger bands* around a 20-day moving average of the price, at the value of +/- 2 standard deviations of the price movement over the last 20 days. The size of the envelope changes over time, depending on the recent volatility of the market. If the next price recorded exceeds the envelope, an entry signal is triggered. The basic idea is to identify when prices change more than would be expected based on their recent volatility, possibly indicating a movement to a new trading range.

The underlying idea of all breakout indicators is simple. All large price moves begin as small price moves, and the breakout indicators attempt to enter the market once a small, but seemingly significant move is detected. Systems based on these indicators are therefore trend-following. The trick is to set the indicators at levels which discriminate well between genuinely trending markets and markets which are just fluctuating randomly (to avoid whipsaw) but not so conservatively that the entry into the trending market is triggered too late to profit from it.

Pairs-Trading

A related application of the breakout idea is found in the *pairs-trading* investment strategy, where the divergence of interest is not between the current and the recent price history of a financial instrument, rather it is concentrated on the ratio of the prices of two separate financial instruments. Therefore, in pairs-trading an investor adopts a relative value investment strategy. Suppose you can identify two companies with similar characteristics, you would expect that their share prices would be correlated as they are exposed to the same environment. If the ratio of the two share prices deviates notably from its historical mean, the expectation of the trading strategy is that the ratio will tend to revert to its long-run average. Hence, the trading strategy is to short the outperformer, and go long on the underperformer. When the ratio moves back towards its long-run average, the two trading positions are closed. In essence, a pairs-trading strategy is a contrarian strategy, whereby the recently (relatively) well-performing share is sold, in favour of a recently (relatively) underperforming share. A trading system based on pairs trading can utilise technical indicators based on a time-series of the price ratio for the pair of shares, for example a MACD indicator based on the ratio.

9.1.4 Stochastic Oscillators

Stochastic oscillators are used to determine when the market is over-bought or over-sold. Generally they compress price data into a fixed range, typically 0-100, hence the name oscillators (as the indicator can only vary between the upper and lower bound). The following provides an example of a simple oscillator:

$$\frac{C - L}{H - L} * 100 \tag{9.3}$$

where C is the current price, L is the lowest price in the last x days and H is the highest price in the last x days. A value approaching 0 is considered to indicate a market which is oversold which will tend to rise. A value approaching 100 indicates a market which is overbought. The task for a human or artificial analyst is to develop filter rules in order to interpret the values of the indicator. For example, critical points may be set at 20 and 80, with values greater than 80 triggering a sell signal and values less than 20 triggering a buy signal. A sample trading rule based on this is:

IF Osc(x) > a THEN Go Short

ELSE IF Osc(x) < b THEN Go Long

where $Osc(x)$ is the value of the oscillator calculated over the last x days, $a > b$, with for example, $x = 12$, $a = 80$ and $b = 20$. Oscillator indicators can be combined with *signal lines* to generate a market entry signal. A signal line is calculated by taking a moving average of the oscillator. If the current value of the oscillator crosses above the signal line, an entry signal (buy signal) is triggered.

RSI Oscillator

Many variants on the basic oscillator exist. Another common version is the *Wilder Relative Strength Indicator* (RSI):

$$100 - \frac{100}{1 + RS} \qquad (9.4)$$

where RS is the average of the daily gains when the share closed up (its closing price was above its opening price) over the last N days/average of the daily losses when the share closed down, over the last N days. A filter rule is developed to generate trading (buy/sell) signals from the RSI. The number of days chosen (N) impacts on the sensitivity of the oscillator. The shorter the time period selected, the more sensitive the indicator.

$\%K$ and $\%D$ Stochastics

Two widely used stochastics are $\%K$ and $\%D$. They are based on the observation that as prices trend upward, closing prices tend to be closer to the upper end of their recent trading range. In down-trending markets, closing prices tend to be closer to the lower end of their recent trading range. This indicator seeks to determine the position of the current closing price relative to its trading range in the last k days. The formula for calculation of $\%K$ is:

$$100 * \frac{\text{Close - Lowest value in past } x \text{ days}}{\text{Highest value in past } x \text{ days - Lowest value in past } x \text{ days}} \qquad (9.5)$$

The indicator is scaled between 0 and 100, with higher values indicating that the closing price is near the top of its recent trading range. A variant on this stochastic is $\%D$ which is a moving average of $\%K$. An example of a trading rule based on these stochastics is:

IF $(\%K < a$ and $\%D < a)$ *AND* $\%D \geq \%K$ *THEN Go Long*

ELSE IF $(\%K > b$ and $\%D > b)$ *AND* $\%D \leq \%K$ *THEN Go Short*

where $a < b$. Stochastics tend to work best in sideways or non-trending markets, and tend to identify small price reversals in relatively flat markets. In strongly trending markets they are less useful, and can become stuck at extreme values at either end of their range while the trend persists.

9.1.5 Volume Data

The same market dynamics that give rise to price also give rise to trading volume. Technical analysts believe that volume is an important technical indicator as volume is considered to precede price [119]. Therefore changes in volume can act as a lead indicator of coming price changes. An intuitive example of this would be the fall-off in buying pressure and trading volume, that occurs as a share nears a price peak (a consensus in the market that the asset is fully valued).

Price and volume information can be combined to obtain a measure of market strength. A market is considered strong by technical analysts if both price and volume are rising. Therefore, a buy signal generated by a trend-following system, perhaps based on moving averages, may be confirmed by comparing current period volume with a lagged moving average of volume. If current trading volume exceeds a lagged moving average of volume, this suggests that other investors are 'participating' in the trend, suggesting that the buy signal is strong.

Ease-of-Movement Indicator

Volume is also closely related to price volatility. The greater the divergence in the opinions of investors as to the true worth of a financial asset, the more likely that the asset will be traded from a pessimistic investor to an optimistic one. One measure which is sometimes used to examine the relationship between price and volume changes is the *Ease-of-Movement* indicator. High (positive) ease of movement values occur when prices are moving upwards on low volume, low (negative) ease-of-movement values occur when prices are moving downwards on light volume. If prices are not moving, or if heavy volume is required to move prices, ease of movement values are close to zero. Generally, the indicator is traded by buying equities when the indicator crosses zero to become positive, and selling when it crosses zero to become negative. The ease-of-movement indicator is calculated using the formula in (9.6), where the mid-point move and the box ratio are calculated as in (9.7) and (9.8).

$$\text{EOM} = \frac{\text{Mid-point move}}{\text{Box Ratio}} \tag{9.6}$$

$$\text{Mid-point move} = \frac{(\text{Today's high} + \text{today's low})}{2} \tag{9.7}$$
$$- \frac{(\text{Yesterday's high} + \text{yesterday's low})}{2}$$

$$\text{Box Ratio} = \frac{\text{Volume (in 10,000s)}}{\text{Today's high - today's low}} \tag{9.8}$$

Trading Volume and Financial Economics

The study of trading volume has also attracted the attention of financial economists. *GARCH* (generalised auto regressive conditional heteroskedacity) models have been applied by financial economists to examine the time-varying nature of stock-price volatility, measured as absolute or squared price changes, usually employing explanatory variables such as trading volume. These models suggest a positive relationship between trading volume and share-price volatility. Another implication of these findings is that if the market is divided into periods of high and low volatility, in the first case price trends persist for shorter time periods than expected (the market constantly switches direction), whereas in periods of low volatility trends in price tend to persist for longer than would be expected if they were truly random.

An analysis of volume data for most financial markets shows that raw volume data tends to be volatile. Trading models incorporating volume data often smooth it by using a moving average of volume, or the rate of change of volume over a recent time period, rather than a raw volume input. Just as for price data, indicators can be constructed to examine the level of momentum in volume data.

9.1.6 Other Indicators

Many other indicators of market sentiment can be included in a trading model, including:

- recent changes in value of market index;
- number of advancing/declining issues as a percentage of all issues; technical analysts suggest that this metric should move in the same direction as general market prices;[2]
- number of shares reaching new highs vs number of shares reaching new lows;
- short-interest ratio;
- volume of options traded;
- ratio of put vs call options traded, and
- the VIX index.[3]

---

[2]A more sophisticated version of this metric is the *Arms Index or TRIN (Trading Index)*, defined as $\left(\dfrac{\text{Number of advancing issues/Number of declining issues}}{\text{Volume of advancing issues/Volume of declining issues}} \right)$, where volume is the number of shares traded.

[3]The CBOE Market Volatility Index (VIX) is a measure of the volatility expectations for the US equity market. It provides investors with up-to-the-minute market estimates of expected volatility by using real-time OEX index option bid/ask quotes. The index is calculated by taking a weighted average of the implied volatilities (the volatility percentage that explains the current market price of an option) of eight OEX calls and puts, which have an average time to maturity of 30 days. Although

As well as constructing technical indicators for the market which it is intended to trade, they can be calculated for time-series of financial data drawn from related markets (intermarket data). Technical indicators are not used in isolation, and analysts will look for confirmation between indicators before relying on them. Trading systems based on technical analysis can, and often will, combine indicators from several financial markets.

9.2 Using Technical Indicators in a Trading System

Technical indicators can be incorporated in two ways as an input in a trading model. Individual indicators can be used directly as model inputs, or, alternatively, indicators can be preprocessed to produce a model input by taking ratios of individual indicators, or through the use of IF-THEN statements (as above). As an example of the latter, a 0 or 1 could be output from a compound rule IF-THEN rule, such as IF (indicator $x > 2$) AND (indicator $y < 4$) THEN 'trading signal' (buy/sell).

Looking at the form of the above IF-THEN rule, it is apparent that a trader will face several subsequent decisions if he decides to use technical indicators as inputs to a trading system:

- Which indicators will be used?
- What parameter values (lag periods/trigger values) should be used?
- How should the indicators be combined to produce a trading signal?

This represents a combinatorial problem. Using traditional modelling methods to determine these choices is likely to prove problematic, as there are effectively an infinite number of possibilities open to the modeller. This suggests that an evolutionary algorithm in which the model structure and model inputs are not fixed a priori will have particular potential for generating trading rules drawn from individual technical indicators.

Technical Indicators as a Regime Detector

Technical indicators can act as a regime detector. For example, consider the expression $RSI_{15}(t) < 0.25$ (a test of whether the RSI over the last 15 days is less than 0.25). If this is true, it indicates that there have been heavy price falls in recent trading days. This 'state of the market' information could be combined with other technical indicators or fundamental/inter-market data to produce a trading signal.

---

the VIX is intended to indicate the implied volatility of 30-day index options, it is used by traders as a general indication of index option implied volatility.

9.3 Summary

The basic premise of technical analysis is that a time-series of price and volume information has information content for the prediction of future stock prices. Four groups of technical indicators are identified, and examination of each group suggests that they will not respond equally well in all market conditions. Each grouping responds best to specific market conditions. If a trader is intending to develop a trading model which is primarily trend-following, technical indicators such as moving average and momentum may be useful. If the intention of the system is to identify short-term tops or bottoms in share prices, then oscillator indicators may be more useful.

As already noted, the usefulness of technical analysis is disputed by many financial economists, who point to studies which have tested simple technical indicators and found that they do not produce excess risk-adjusted returns. A shortcoming of these studies is that they do not test technical analysis as it is actually used by investors. Successful technical analysts do not just apply simple rules drawn from basic technical indicators. Rather they use rules which consist of multiple indicators, which must agree before action is taken by the investor. In addition, successful analysts employ filter rules to determine which trading rules should be applied in specific market conditions.

Readers who require further detail on technical analysis are referred to the many good specialist texts which exist on this topic [119, 122, 160].

Part III

Case Studies

10

Overview of Case Studies

The case studies in the following ten chapters illustrate how the biologically inspired methodologies discussed in earlier chapters can be applied to real-world financial applications. Each case is concluded with a few suggestions as to how the developed model could be extended. A brief synopsis of each case follows.

The first case study is an application of a basic feedforward MLP to construct a financial prediction model. The second case provides an example of a hybrid GA-MLP methodology. The GA is employed to evolve the weight vectors and network structure, but could easily be adapted to also select the model's inputs and choice of transfer functions. The third case demonstrates the use of GE to evolve a basic trading system, using technical indicators as inputs. A particular utility of the GE methodology is that it can be used to automate both the hypothesis generation and hypothesis optimisation steps. The approach also allows the evolution of trading models with investor-desired risk/return characteristics. The fourth case demonstrates an extension of the third case by demonstrating how GE can be combined with a moving-window approach, to create a real-time adaptive trading model. The fifth case also extends the third case, using high-frequency data, and by illustrating the importance of choosing exit strategies in a trading system carefully. The sixth case demonstrates how GE can be used to evolve a trading system for spot foreign exchange markets.

The final four cases demonstrate the broad utility of biologically inspired methodologies concentrating on the prediction of corporate failure and corporate bond ratings. The seventh and eighth cases provide an example of how GE and an ant colony-inspired algorithm, respectively, can be applied to construct a model that anticipates corporate failure. The ninth and tenth cases demonstrate how GE and AIS can be used to evolve models for corporate bond rating.

A number of the cases in the following chapters have been co-authored, and we wish to acknowledge the contribution of our co-authors. Ian Dempsey co-authored the adaptive trading with GE chapter (Chap. 14). Peter Keenan,

Katrina Meagher and Edward Carty contributed to the development of the intra-day trading case study (Chap. 15). Yue Xi and Qiang Han contributed to the ant-model corporate failure case (Chap. 18). Finally, Peter Keenan, Alice Delahunty and Denis O'Callaghan contributed to the AIS bond-rating chapter (Chap. 20).

11

Index Prediction Using MLPs

A market index is comprised of a weighted average measure of the prices of the individual shares which make up that market index. Movements in a market index are therefore indicative of changes in the balance of supply and demand for the individual shares making up the index. Apart from acting as a general barometer of the market, the values of a variety of exchange-traded funds (ETFs)[1] and a number of futures and options products are tied to the values of market indices. Hence, the ability to even partially forecast the future value of an index would be of considerable interest to investors trading these products. This case study examines the ability of a series of multi-layer perceptrons (MLPs) to predict the five-day percentage change in the value of the UK's FTSE 100 index during the period June 1995 to December 1996, using a combination of technical, fundamental and intermarket data.

Figures 11.1 and 11.2 provide two graphical perspectives of the five-day percentage change in the value of the FTSE 100 index. Figure 11.1 suggests that while most 5-day changes are within a range of $-/+2\%$ percent, larger changes are not uncommon. Large swings occur quite a bit more often than would be expected if they followed a normal distribution. Figure 11.2 displays the changes in a histogram format, and includes a normal curve for reference purposes. The histogram of the five-day percentage change displays a *lep-*

---

[1] An ETF aims to mirror the performance of a particular market index. Therefore, an investor aiming to produce a return which is tied to the performance of a market index need not buy a basket of individual shares, but can instead buy an ETF product. ETFs are themselves quoted on stock markets and can be bought and sold just like shares of companies. ETFs go under a variety of unusual names like *Qubes* which mimic the performance of the Nasdaq 100 index (ticker symbol QQQ), *Diamonds* which mimic the performance of the Dow Jones Industrial Average, and *Spiders* which mimic the performance of a number of S&P indices. In addition to ETF products which track broad market indices, there are also ETFs which track the performance of sectoral indices. ETF products are available for most large stock exchanges. Traders can also 'short' ETFs and buy them on margin.

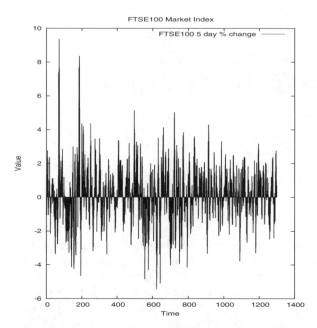

Fig. 11.1. A plot of the five-day % change in the value of the FTSE 100 over the period 1/1/92 to 31/12/96. The y-axis represents the % change in the index over a five-day period

tokurtotic distribution, which is a general characteristic of time-series of price changes of equities. Leptokurtotic or *heavy-tail* distributions exhibit more frequent large positive and negative price changes than would be expected if price changes followed a normal distribution. Distributions of financial data also exhibit higher 'peaks' than would be expected if they followed a normal distribution. Distributions with these two properties are sometimes referred to as *stable Paretian* or *fractal* distributions.

11.1 Methodology

The predictive horizon, that is the length of time of the financial forecast, plays a significant role in determining the utility of potential input variables in index prediction. For example, many macroeconomic variables that would be expected to play a role in determining stock prices are reported infrequently and are predominantly invariant in the context of a short-term prediction model. Future expectations of these variables are not invariant, but by their very nature are difficult to observe!

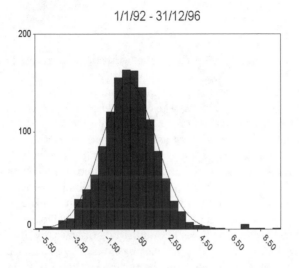

5 day percentage change in FTSE 100 Index

1/1/92 - 31/12/96

Fig. 11.2. A histogram of the five-day % changes in the value of the FTSE 100 over the period 1/1/92 to 31/12/96. The x-axis is denominated as the % change in the index over a five day period

Input Selection

The inputs used in the model in this case study were selected from a range of technical, fundamental and intermarket data suggested in prior literature [35, 37, 63, 160]. Initially, a series of technical indicators drawn from price, volume and options data were calculated. These included a variety of moving averages, relative strength indicators, oscillators and lagged measures. Chap. 9 provides an introduction to technical indicators and technical analysis. An infinite number of technical indicators could be calculated based on historic index data. In this case we restricted attention to indicators calculated within a 20 day period prior to the forecast period.

In order to select the final subset of technical indicators used as inputs to the model, several data selection tools were employed, including correlation analysis, the construction of regression models and the construction of preliminary MLPs. A similar selection process was applied to select the intermarket and fundamental indicators. Ten inputs were included in the final model:

  i. 5-day lagged percentage change in the value of the FTSE 100 index
 ii. 20-day lagged percentage change in the value of the FTSE 100 index
iii. Ratio of the 10 vs 5-day moving average of the value of the FTSE 100 index
 iv. Ratio of the 20 vs 10-day moving average of the value of the FTSE 100 index
  v. Bank of England Sterling index

vi. S&P 500 composite index$_{(t)-(t-5)}$
vii. LIBOR 1-month deposit rate
viii. LIBOR 1-year deposit rate
ix. Aluminium ($ per tonne)
x. Oil ($ per barrel)

The first two inputs provide the model with a measure of the 'momentum' in the market, and also with the ability to discern whether a short-run 5-day trend in the market index agrees with the longer, 20-day trend. The next two inputs calculate two moving average convergence-divergence (MACD) metrics of differing length. A measure of the BoE Sterling index is included as changes in the Sterling exchange rate against major trading partners can be expected to impact on the domestic and overseas earnings of firms in the FTSE 100 index. For similar reasons, the model is provided with inputs on the major raw materials oil and aluminium. Changes in exchange rates and commodity prices also provide a leading indicator of inflation rates. Finally, two measures of interest rates (London inter-bank rates) are provided, each of differing term. Interest rates affect share prices in a multitude of ways, by altering the rate of return which can be earned on competing financial assets such as bonds and bank deposits, by their impact on the borrowing costs of firms, and by their impact on the general macroeconomic climate. To satisfy the constraints of the MLP model, and to facilitate learning, all inputs were normalised into the range (-1,1).

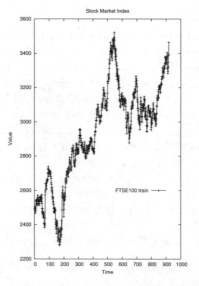

Fig. 11.3. A plot of the FTSE 100 over the training period

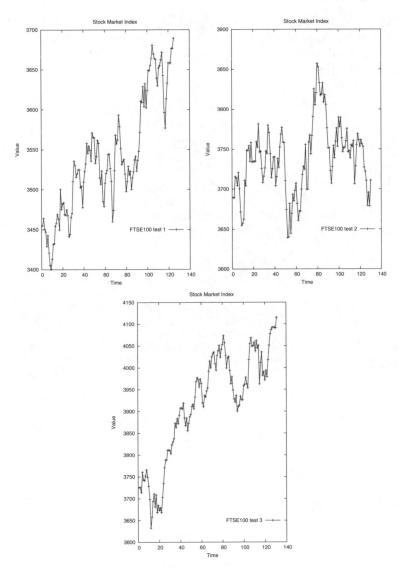

Fig. 11.4. A plot of the FTSE 100 over each of the three out-of-sample test periods

Dataset

The data used in model development and testing was drawn from the period 1/1/92 to 31/12/96. Each model was developed using 918 days of trading data. A total of 793 days data was used to train each model. To reduce the problem of overtraining, model performance on the remaining 125 days of data (the validation dataset) was used to select the best model. The models were tested out of sample on three subsequent six-month periods (10/7/95-31/12/95, 1/1/96-

28/6/96 and 1/7/96-31/12/96). Figure 11.3 provides a graph of the FTSE 100 index over each of the train/test periods. Testing each six-month period separately facilitated the examination of the evolution of the predictive performance over time.

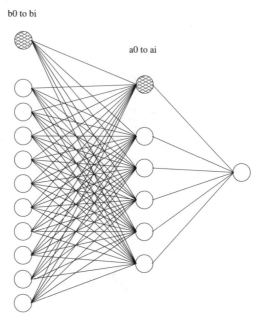

Fig. 11.5. The MLP model adopted for this case study has 11 input nodes, 6 hidden nodes, and a single output node

11.1.1 Model Selection

In developing the final MLP models, a fixed 11:6:1 structure was utilised (see Fig. 11.5):

$$y_t = L \left(\sum_{j=0}^{5} w_j L \left(\sum_{i=0}^{10} b_i w_{ij} \right) \right) \tag{11.1}$$

where b_i represents $input_i$ (b_0 is a bias node), w_{ij} represents the weight between input node$_i$ and hidden node$_j$, w_j represents the weight between hidden node$_j$ and the output node, and L represents the hyperbolic tangent function. The hyperbolic tangent function has the property of non-linearly squashing inputs into the range [-1,+1], and can be seen in Fig. 11.6. The size of the hidden layer was fixed *a priori* at six hidden nodes. The choice of the tanh transfer function is governed by the requirement that the value of

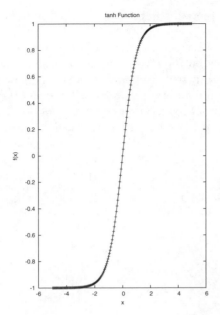

Fig. 11.6. The hyberbolic tangent function

the index could fall as well as rise, hence the model must be able to output a
negative result.

Postprocessing the Output

Examining the typical percentage changes in the index over a five-day period,
we note that it rarely changes by more than +/- 6%. We can therefore rescale
the outputs in the training data into a [-1,+1] range by dividing them by
a factor of ten. Once the model is trained, the predicted output is rescaled
back from the range [-1,+1] by multiplying it by ten, to produce the forecast
percentage change in the index for the next five days. The choice of six nodes
in the hidden layer was arrived at after initial experimentation with differing
numbers of hidden layer nodes. The predictive accuracy of the network was
found to be fairly insensitive to a choice between five and seven hidden layer
nodes, and the final models were developed with six hidden layer nodes.

11.1.2 Model Stacking

Most reported applications of an MLP modelling methodology to index prediction consist of the construction of a single model. Given the limitations
of a problem domain in which input/output relationships are dynamic and
where input data is incomplete and noisy, no single model may be dominant

and there may be potential to improve predictive quality by building multiple models [26].

In this case, we construct 25 separate MLP models using both different initial starting points on the error surface (by using different weight vector initialisations)[2] and different randomisations of data between training and (in-sample) validation datasets. This approach reduces the dependence of the final prediction on initial conditions stemming from the order of input/output data vectors in the dataset used for training and validation purposes.

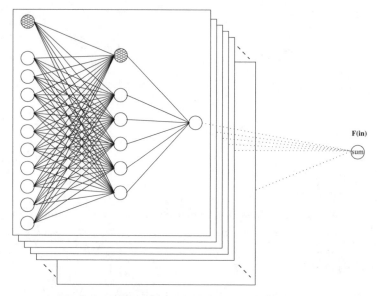

Fig. 11.7. The MLP model adopted for this case study has 11 input nodes, 6 hidden nodes, and a single output node, and 25 separate MLP models are generated and their outputs combined by committee to determine a trading signal

In a similar concept to the combination of the output of simple non-linear processing elements in an individual MLP, the predictions of individual MLPs are combined (*stacked*) to form a *committee decision*. The overall output of the stacked network of MLPs is a weighted combination of the individual network outputs:

$$F(in) = \sum_{i=1}^{n} w_i f_i(in) \tag{11.2}$$

---

[2]The 'bad initialisation' problems which can arise in MLPs are reduced when using evolutionary algorithms. Evolutionary algorithms explicitly maintain a population of potential solutions, rather than attempting to iteratively improve a single solution, and therefore are much less susceptible to poor choices of initial starting point

where F*(in)* is the output from the stacked network for input vector *(in)*, $f_i(in)$ represents the output of *network$_i$*, and w_i represents the stacking weight. In this example, the predictions are combined on an equally weighted, linear basis. Thus, in calculating the predicted 5-day percentage change in the market index, the average prediction of all 25 models is used. Figure 11.7 provides an overview of the adopted model.

The predictive ability of an individual MLP is critically impacted by the choice of network weights. Since the back-propagation algorithm is a local search technique, the initial weights can have a significant impact on the solution chosen by the neural net. The stacking process is introduced both to reduce this problem and to combine the predictive abilities of individual models, which may possess differing pattern recognition capabilities. The decision to select 25 models for stacking purposes is guided by the findings of Zhang et al. [225], which suggest that stacking 20 to 30 networks is usually sufficient to stabilise model errors.

11.2 Results

This section outlines the results obtained in each of the out-of-sample test periods, and discusses the robustness of the models' predictions over time. A simple trading system is also developed based on the models' prediction, and the performance of this system is benchmarked against that of a simple buy-and-hold strategy.

11.2.1 RMSE and Correlation

Table 11.1 summarises the results for each out-of-sample test period, showing the average RMSE, calculated as the average of the RMSEs for each of the 25 model's individual daily predictions. This is contrasted with the RMSE of the average prediction of all 25 models taken together.

Table 11.1. Average RMSE of each individual model's predictions, and of the combined prediction of all 25 stacked MLPs

	Test set 1	Test set 2	Test set 3
Average of Individual Models	0.2335	0.2754	0.3598
Stacked & Combined	0.1833	0.1736	0.2143

The averaging of predictions across 25 networks has resulted in a noticeably lower error. It is also noteworthy that the RMSE has generally increased over the three time periods, indicating that the predictive quality of the models is degrading over time. To investigate this point further, the correlation co-efficients between the model's (models') predicted output and the actual percentage five-day index change were calculated. These are summarised in Table 11.2.

Table 11.2. Average correlation actual vs predicted 5-day change averaged over the individual models, and compared against the combined prediction of all 25 stacked MLPs

	Test set 1		Test set 2		Test set 3	
	Pearson	Spearman	Pearson	Spearman	Pearson	Spearman
Average of Individual Models (r)	0.2830	0.2833	0.2009	0.2217	0.1669	0.1689
Stacked (r)	0.3973	0.4049	0.2902	0.3028	0.2231	0.2293

The predictions made as a result of averaging the predictions over 25 models show a higher correlation to the actual five-day change than do the average correlations of the individual models. In both cases, the correlation coefficients show degradation over time. For comparison purposes, a linear regression model that used the same inputs as the MLPs was fitted to the training dataset. When applied to the first out-of-sample dataset, it resulted in a RMSE of 0.7417, and a Pearson correlation between the predicted and the actual five-day change in the index of 0.0023, indicating the inability of the linear regression model to usefully anticipate the five-day change in the index value.

Model Specification

The degree of bias of the predicted values produced by the MLPs can be assessed by regressing the predicted against the actual five day percentage changes for the index [184]. For an unbiased forecast:

$$y(actual) = a + b(forecast) + u \qquad (11.3)$$

where $a = 0, b = 1$ and u is white noise. The regression of the predicted values against actual across the out-of-sample test periods produces, $y = 0.14451 + 0.4851(forecast)$ with standard errors for a and b of 0.0610 and 0.0686. In this case, the estimated regression co-efficients are significantly different (at the 5% level) from 0 and 1 respectively, which leads to a rejection of an unbiasedness hypothesis. This finding is consistent with a possible mis-specification of the model and with the intuition that no relatively small model is going to capture more than a portion of the complexity in the data-generation processes of financial markets. Hence, we should remember that any models of markets we construct are simplified representations of the complex, underlying data-generating process. No static model will work well across changing market conditions. This suggests that better models will have an adaptation mechanism. A simple means of adaptation is to retrain the model using new data drawn from current market conditions, or to weight recent data more heavily in model training. More complex adaptive mechanisms include embedding an adaptive capacity in the modelling methodology itself, for example by allowing the system to alter the inputs it uses in producing its predictions, over time. As will be seen in following chapters, one particular advantage of evolutionary approaches is that they can automate this step.

11.2.2 Trading System

A simple trading system was developed based on the models' predictions. The performance of this system was benchmarked against that of a simple buy-and-hold strategy. The trading strategy adopted was to invest $1,000 each time the average of the models' predictions of the five-day percentage change exceeded $|1.5\%|$. This position was automatically closed five trading days later. The actual investment at a given point in time is not determinable ex ante, and this makes it difficult to perfectly determine the size of an appropriate, equivalent risk, buy-and-hold investment. The approach adopted is to calculate the buy-and-hold investment using the average daily ex post investment arising under the MLP-driven trading strategy. Interest costs and dividends are ignored. The comparative results of each investment strategy, ignoring trading costs, are shown in Table 11.3. The comparative results when a 1% trading cost is included (allowance for commission and slippage) are shown in Table 11.4. Table 11.3 suggests that in the absence of trading costs it may be possible to develop a profitable trading system based on the predictions of the neural models.

Table 11.3. Comparative results of investment strategies ignoring trading costs

Trade if $> |1.5\%|$	Test set 1	Test set 2	Test set 3
Profit	$130.72	$159.45	$0
Return on investment (%)	40.85	20.73	0
Win ratio (%)	100	75	n/a
Buy-and-hold profit	$21.70	$4.54	$0
Buy-and-hold profit (%)	6.78	0.59	n/a

The degradation of the usefulness of the trading system over the three test periods is consistent with a hypothesis that market structure is dynamic. When trading costs are included in the analysis (Table 11.4), it is unclear whether a trading system based on the models' predictions outperforms a buy-and-hold strategy, as the trading system shows superior performance in the first test period and poorer performance in the second. As with all trading systems developed and back-tested using historical price data, a number of

Table 11.4. Comparative results of investment strategies including trading costs

Trade if > |1.5%|	Test set 1	Test set 2	Test set 3
Profit	$50.73	-$40.55	$0
Return on investment (%)	15.85	-5.27	0
Win ratio (%)	75	50	n/a
Buy-and-hold profit	$18.50	-$3.15	$0
Buy-and-hold profit (%)	5.78	-0.41	n/a

caveats must be borne in mind. Live markets have attendant problems of delay in executing trades, illiquidity, interrupted/corrupted data and interrupted markets. The impact of these issues is to raise trading costs and consequently to reduce trading profitability.

11.3 Discussion

The aim of this case was to illustrate an application of a MLP to predict the five-day percentage change in the value of the FTSE 100 Index. The results suggest that neural network models can be constructed which have predictive ability, that the structure detected by the models is persistent, and that in the absence of a major market shock predictive quality degrades gracefully. The results also suggest that a stacking methodology can improve predictive quality. A simple trading system, developed using the networks' predictions outperformed a simple buy-and-hold strategy when trading costs were ignored, but when trading costs were included the results were ambiguous. This case study did not attempt to develop an optimal trading system and the results provided should be considered a lower bound on those which could be achieved from a trading system based on the trained MLP.

Extending the Model

Extensions of this basic model would include the construction of a more sophisticated set of entry and exit strategies, and the testing of the utility of a wider range of possible model inputs. In addition, it would be interesting to investigate a number of alternative strategies to combine the 25 stacked MLP models. For example, with a temporal adaption of the weights associated with each layer of the stack, or by using an evolutionary automatic programming programming approach, such as genetic programming or grammatical evolution, to generate the combination function itself.

Many, more sophisticated targets other than price change over the next x days could have been used for the MLP. Rather than attempting to predict price changes, the MLP could be constructed to predict market 'turning points' using oscillators as inputs (with the aim of buying when a market bottom is indicated, and buying when a market top is indicated), or to predict forthcoming channel breakouts (with the aim of buying if an upward breakout occurs, and selling if it is anticipated that a support level will be breached). Another possibility is to develop an MLP to predict market turning points, by detecting the divergence of markets which are fundamentally linked.

The case could also be extended by using a moving window approach when training the MLP (see Chap. 14) where the system only predicts one step (or a small number of steps ahead) at a time. The model is then continually retrained as new data becomes available.

Opportunities for Using an Evolutionary Methodology

As can be seen in this case a great deal of effort was invested into the selection and generation of suitable model inputs, which in some cases required and profited from the use of domain knowledge. In the event that knowledge is unavailable, or in the worst case may not exist (yet), a method that can automatically and adaptively select appropriate input variables could be beneficial. We will see in a later study (Chap. 17) how an evolutionary automatic programming approach can provide a solution in this scenario.

The next chapter extends the above case study by illustrating an approach to evolve both the topology and weights of a MLP, therefore partly automating the construction of the MLP.

Index Prediction Using a MLP-GA Hybrid

Applications of NNs in business and finance are generally developed through a trial and error approach, guided by heuristics. This process is time-consuming, and there is no guarantee that the final network structure is optimal. The objective of this case is to illustrate how an evolutionary algorithm, the genetic algorithm (GA), can be utilised to develop both the connection structure and connection weights for a MLP. The case examines the predictive quality of the resulting MLP by comparing it with a benchmark MLP consisting of a three-layer, fully connected, feedforward structure, trained using the back-propagation training algorithm.

12.1 Methodology

The same input data is used as in the last case study, and again the objective is to predict the five-day percentage change in the value of the FTSE 100 index. The inputs used were:

   i. Five-day lagged percentage change in the value of the FTSE 100 index
  ii. Twenty-day lagged percentage change in the value of the FTSE 100 index
 iii. Ratio of the five vs ten-day moving average of the value of the FTSE 100 index
  iv. Ratio of the ten vs twenty-day moving average of the value of the FTSE 100 index
   v. Bank of England Sterling index
  vi. S&P $500_{(t)-(t-5)}$ composite index
 vii. LIBOR one-month deposit rate
viii. LIBOR one-year deposit rate
  ix. Aluminium ($ per tonne)
   x. Oil ($ per barrel)

As in the last case, all inputs were normalised into the range (-1,1), as was the predictive target.

12.1.1 Model Construction

In constructing the MLPs several factors were held constant. The same dataset was used for training and out-of-sample validation, and the network structure was fixed as a three-layer feedforward network with a hyperbolic tan transfer function. This reduced the benchmark MLP development process to a decision regarding the number of nodes in the hidden layer.

In the hybrid MLP-GA, both the number of active connections between the input→hidden and the hidden→output layers, as well as the values of the connection weights were determined using the GA (Fig. 12.1). Initially, all connections are set to 1 (connected), which ensures that the search process starts with a fully connected network. Implicitly, by allowing the GA to evolve the number of connections, the process also determines the number of hidden nodes in the resulting MLP. A hidden layer node with no active output connections is effectively pruned from the network, as is a hidden layer node with no active incoming connections.

Representing the MLP

As both the connection and weight values are evolved, the strings representing each individual MLP structure in the population acted on by the GA must contain the information required to decode both the weights attached to each connection in a network and a binary indicator which indicates whether each connection is active in a network. In order to bound the search space, the weight values were limited to a range between -0.8 and +0.8. This constraint also helps reduce the chance that the evolutionary process will produce a forced model with extreme weight values. In addition, bounds were set on the maximum number of hidden nodes in the evolved MLP.

The structure of connections in each MLP can be represented as a connection matrix, consisting of 0s and 1s, where a 1 in position (x, y) indicates that node x is connected to node y (Fig. 12.2A). The matrix can also be represented in a linear genome form by concatenating each of its rows. The real-valued weights corresponding to each of the connections are placed on the string in connection order, immediately after the corresponding row of the connection matrix. This representation ensures that the connectivity variables (off or on) for connections into each node and the associated weights for each of these connections are co-located on the binary string.

The Evolutionary Process

A population of 150 mixed-form genotypes was initially randomly generated. An evolutionary process was then applied to this starting population. In order to select pairs of model structures for reproduction, a rank order procedure was adopted. This method ranks all solution encodings such that encoding i is ranked ahead of encoding j when $f(i) > f(j)$, where $f(i)$ is the fitness

b0 to bi

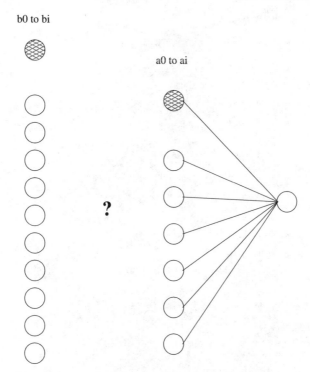

Fig. 12.1. Both connections and weights are evolved to nodes in the hidden layer

of encoding i. One encoding is fitter than another when it produces a lower RMSE on the training dataset.

A selection function derived from a negative exponential function was used to calculate $p(i)$, which represents the probability that encoding i is selected for reproduction. This biases the selection process towards the highest ranking solution encodings. The selection function is constrained to ensure that $\sum_i p(i) = 1$. An *election* operator was also employed to ensure that the current best individual in the population was copied unchanged into the next generation of encodings.

Single-point crossover was applied with a probability of 0.5, the crossover point being randomly selected from the end points of the connection 'weight blocks' on the genotype. A simple mutation operator was applied to each element of the GA chromosome with a probability of 0.05. Binary elements on the chromosome were flipped 0 to 1 or vice versa, and real-valued elements on the chromosome were mutated by adding a randomly drawn value within the range [-0.1,+0.1] to their current value, and limiting the resulting value to the range [-0.8,+0.8]. After the selection, crossover and mutation processes, the current generation of encodings was replaced by the newly generated en-

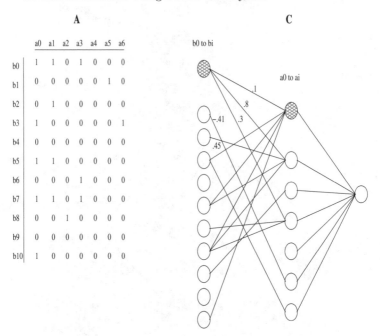

A

	a0	a1	a2	a3	a4	a5	a6
b0	1	1	0	1	0	0	0
b1	0	0	0	0	0	1	0
b2	0	1	0	0	0	0	0
b3	1	0	0	0	0	0	1
b4	0	0	0	0	0	0	0
b5	1	1	0	0	0	0	0
b6	0	0	0	1	0	0	0
b7	1	1	0	1	0	0	0
b8	0	0	1	0	0	0	0
b9	0	0	0	0	0	0	0
b10	1	0	0	0	0	0	0

C

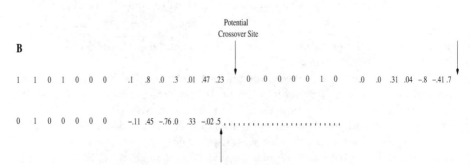

B

Potential
Crossover Site

1 1 0 1 0 0 0   .1 .8 .0 .3 .01 .47 .23   0 0 0 0 0 1 0   .0 .0 .31 .04 −.8 −.41 .7

0 1 0 0 0 0 0   −.11 .45 −.76 .0   .33 −.02 .5 ,,,,,,,,,,,,,,,,,,,,,,,,,

Fig. 12.2. Representation of the MLP as a connection matrix (A), the GA chromosome (B), and the final MLP model (C)

codings, adopting a standard generational replacement strategy with elitism, as the best encoding is always kept.

12.2 Results

The benchmark MLP was set as an 11-7-1 structure, comprising of the ten inputs, six hidden layer nodes, and a bias node in both the input and hidden layers. This results in a model with 84 weights. The results from this model are shown in Table 12.1.

Table 12.1. Correlation coefficients and coefficients of determination (between actual and predicted outputs) for the fully connected (11-7-1), feedforward network

	Training data	Test set 1	Test set 2	Test set 3
r	0.2910	0.2937	0.2899	0.0945
R^2	0.0846	0.0862	0.0840	0.0089

The training/validation period consists of 918 data vectors drawn from 1/1/92 to 7/7/95. The out-of-sample dataset is split into three divisions of similar size, to determine whether predictive accuracy declines markedly as the length of time from the training/validation dataset increases. The first out-of-sample dataset consists of 125 data vectors drawn from the period 10/7/95 to 31/12/95. The second and third datasets are comprised of 130 and 132 data vectors, respectively and are drawn from the periods 1/1/96 to 28/6/96 and 1/7/96 to 31/12/96. Out-of-sample predictive performance in the first two test datasets is generally consistent with the model's performance on the training/validation dataset but declines in the third period, indicating that the MLP model requires retraining.

12.2.1 MLP-GA

The MLP-GA modelling approach does not construct a single MLP, but rather constructs a population of MLP structures. An initial population of 150 MLPs was generated randomly, and the evolutionary algorithm was run for 1,500 generations. The evolutionary process was terminated at this point. An examination of the population of evolved MLPs showed that a substantial number had achieved a similar fitness level. The results from the network with highest fitness are shown in Table 12.2.

Table 12.2. Correlation coefficients and coefficients of determination (between actual and predicted outputs) for the highest-fitness MLP evolved

	Training data	Test set 1	Test set 2	Test set 3
r	0.4022	0.4453	0.3239	0.1805
R^2	0.1617	0.1983	0.1049	0.0326

The results present a similar picture to those obtained from the fully connected, feedforward MLP. The predictive quality of the model degrades over the three out-of-sample test periods. The r values for the MLP constructed using a GA are noticeably better than those obtained from the fully connected, feedforward MLP. One notable additional distinction between the models concerns the number of weights (connections) utilised in the MLP-GA hybrid. This model utilises a structure containing 26 connections, less than half the number utilised in the fully connected model. Table 12.3 shows the network structure of the highest fitness member of the population of MLP-GA hybrid models. A 1 indicates that a connection is used, a 0 indicates that it is not (the bias node in the input layer was explicitly connected to all hidden layer nodes and the bias node in the hidden layer was connected to the output node only). In addition, the connection pruning implicitly selects the number of hidden nodes. Only hidden nodes connected to both the input and output layers are 'active' in the final evolved network. As can be seen from Table 12.3, hidden nodes 1, 4 and 5 have no connection to the output node and hence are not included in the final network.

Table 12.3. Best population member MLP-GA structure

| | Hidden Nodes | | | | | |
---	1	2	3	4	5	6
5-day lagged change in FTSE	0	0	1	0	1	1
20-day lagged change in FTSE	0	1	0	1	0	1
5 vs 10-day MA of the FTSE	0	1	1	1	0	0
10 vs 20-day MA of FTSE	1	0	1	1	1	1
BoE sterling index	0	1	1	1	1	1
S&P $500_{(t-1)-(t-5)}$	0	0	0	1	0	0
LIBOR one-month rate	0	1	1	1	1	1
LIBOR one-year rate	1	1	1	1	0	1
Aluminium ($ per tonne)	1	0	0	0	0	1
Oil ($ per barrel)	1	0	1	1	1	0
Output Node	0	1	1	0	0	1

Despite using a sparse structure, the predictive accuracy of the model both in and out of sample as measured using r is greater than that of the fully connected model. This suggests that there is a substantial level of redundancy in the structure of the fully connected model although it should be noted that the structure of the benchmark model was determined as appearing optimal after trial and error experimentation with alternative hidden layer sizes.

12.2.2 Analysis of Weight Vectors

Although decomposition and meaningful interpretation of the weight vectors of a large MLP structure is not a trivial task, several methods exist which can be employed to provide some insight into the workings of the MLP. In order to examine the relative importance of each of the inputs, a *contribution analysis* was performed. The contribution values are calculated to provide a guide to the influence of each input and hidden layer node on the model. The contribution of an input is calculated as the ratio of the absolute values of all the connection weights between that input and the hidden layer and the total of all connection weights between all input nodes and the hidden layer.

Table 12.4 provides the weight matrix and contribution totals for the best MLP-GA hybrid. In constructing this table, the incoming/outgoing weights associated with hidden layer nodes and those which the GA has pruned from the network, as well as the weights of other connections not utilised by the network, are set to zero.

Table 12.4. Weight structure and input node contribution

| | Hidden Nodes | | | Contribution |
	2	3	6	%
Bias to Hidden	-0.2133	-0.8	0.8	14.08
5-day lagged change in FTSE	0.00	-0.1859	-0.114	2.33
20-day lagged change in FTSE	-0.799	0.00	-0.5958	10.82
5 vs 10-day MA of the FTSE	-0.8	-0.7098	0.00	11.72
10 vs 20-day MA of FTSE	0.00	-0.8	0.3182	8.68
BoE sterling index	-0.8	0.8	-0.7982	18.62
S&P $500_{(t-1)-(t-5)}$	0.00	0.00	0.00	0.00
LIBOR one-month rate	0.6318	0.7971	-0.075	11.67
LIBOR one-year rate	0.799	-0.4	0.6827	14.61
Aluminium ($ per tonne)	0.00	0.00	0.8	6.21
Oil ($ per barrel)	0.00	-0.1628	0.00	1.26
Output Node	0.7997	-0.8	-0.5332	100

Although this technique has limitations, broadly speaking the larger the contribution value for an input node the greater the apparent importance of that input. A contribution value can also be calculated for the hidden layer nodes to reveal whether any hidden node is dominating the model. Examining the contribution values, we see that all inputs with the exception of the S&P 500 information are utilised by the model. Looking at the contribution values for the individual inputs we note that they are highest for the Bank of England index, LIBOR, the 20-day lagged change in the FTSE 100 index and the ratio of the 5/10-day moving average of the FTSE 100 index. However, extreme care should be taken when trying to interpret contribution values. The significance of a contribution value depends on several factors, including the method of normalisation of the underlying data series, and the way the data series was preprocessed. For example, consider a data series where most values are close to 0.1, but which has one value of 10. If this data series is simply rescaled into a [0,1] range by dividing by ten, most values will end up close to zero. Hence, even if the weights (and therefore the contribution value) associated with that input are relatively large, it may have little practical impact on the output of the network.

12.3 Discussion

The objective of this case was to illustrate how the global search properties of a GA could be applied to ascertain both a connection structure and the associated weights for a MLP. The hybrid methodology contains both a hypothesis generation component (the GA) and a hypothesis optimisation component (the MLP).

Utilising a series of fundamental and technical market data drawn from the FTSE 100 index as a test bed, both the evolved and the benchmark MLPs attempted to predict the five-day percentage change in the value of the FTSE 100 index. The results suggest that the MLP-GA hybrid model utilised a sparse internal structure but, despite this, outperformed the fully connected benchmark MLP. This would suggest that there may be scope for utilising MLP-GA hybrid combinations in applications where there is a shortage of data for model building and testing.

Scope exists to further develop the basic MLP-GA hybrid outlined in this case. The evolutionary process could be extended to encompass the selection of inputs, the form of transfer functions, and the number of hidden layers. The methodology could also be extended by applying more sophisticated versions of the GA. The GA variant adopted in this case is closely related to the original algorithm outlined by Holland [108]. While the canonical GA is useful in this case as an illustration of a hybrid MLP-GA approach, more powerful alternatives now exist in the form of competent GAs, which demonstrate a superior ability to scale to harder problems than does the simple GA. These competent algorithms model and exploit linkages that exist between the genes of an individual to perform a more effective search by modelling and respecting building blocks. As already described in Chap. 3, efficient search of MLP structures by a GA depends on a good linkage between the representation of the MLP and the definition of the variety generation operators. In this case, we have not attempted to optimise the design of the crossover operator. Yao [203] and [224] provide a good discussion of methods that can be applied for this purpose.

13

Index Trading Using Grammatical Evolution

This case provides an illustration of the application of grammatical evolution to construct a simple trading system based on technical indicators for three market indices, the UK FTSE, the Japanese Nikkei, and the German DAX [171].

As noted by Iba and Nikolaev [114], there are a number of reasons to suppose that the use of an evolutionary automatic programming (EAP) approach such as GE can prove fruitful in the financial prediction domain. EAP methodologies can simultaneously evolve both a good selection of model inputs and a good model form. EAP methodologies also facilitate the use of complex fitness functions including discontinuous, non-differentiable functions. This is of particular importance in the financial domain as the fitness criterion may be complex, usually requiring a balancing of return and risk. Another useful feature of EAP is that it produces human-readable rules that have the potential to enhance understanding of the problem domain. Another advantage of EAP systems is that they allow for the easy construction of complex entry/exit rules (Chap. 15).

13.1 Methodology

The FTSE data is drawn from the period 26/04/1984 to 4/12/1997 and represents the closing value of the index for each day during this period. The training data set was composed of 440 days, and the remaining data is divided into five hold-out samples totaling 2125 trading days. A graph of the FTSE index over these periods is provided in Fig. 13.2. The DAX and Nikkei data is drawn from the period 1/1/1991 to 3/12/1997, and two hold-out samples are used in both cases.

Graphs of each of these market indices over these periods can be seen in Figs. 13.1 and 13.3. The division of the hold-out period into a number of segments is undertaken to allow comparison of the out-of-sample results across different market conditions, in order to assess the stability and degradation characteristics of the developed models' predictions.

Input Selection

This case restricts attention to technical indicators. As outlined in Chap. 9, technical indicators can be broadly grouped into four categories:

   i. moving average indicators,
  ii. momentum indicators,
 iii. range indicators, and
  iv. oscillators.

In the creation of a trading system using technical indicators, the challenge is to select indicators, their associated parameters, and to combine the indicators to produce a trading signal. Given the large search space that this produces, and the impossibility of enumeratively trying all possible combinations, a methodology such as GE has particular promise. In this case study, the first three of the above groupings of indicators are defined in the grammar we use to evolve the trading systems.

Trading Rules

The rules evolved by GE are used to generate one of three signals for each day of the training or test periods: *Buy, Sell* or *Do Nothing*. A variant on the trading methodology developed in Brock, Lakonishok and LeBaron [35] is then applied. If a buy signal is indicated, a fixed investment of $1,000 is made in the market index. This position is closed at the end of a fixed ten day period. On the production of a sell signal, an investment of $1,000 is sold short and again this position is closed out after a ten-day period. This gives rise to a maximum potential investment of $10,000 at any point in time (the potential loss on individual short sales is in theory infinite but in practice is unlikely to exceed the investment of $1,000). The profit (or loss) on each transaction is calculated taking into account a one-way trading cost of 0.2% and allowing a further 0.3% for slippage. To allow comparison of the returns generated by the trading system with those of a buy-and-hold investment strategy, the total return generated by the developed trading system is a combination of its trading return and the risk-free rate of return generated on uncommitted funds. As an approximation this is calculated using the average interest rate over the entire dataset.

The values of the market indices changed substantially over the training and testing periods. Before the trading rules were constructed, these values were normalised using a two-phase preprocessing. Initially the daily values were transformed by dividing them by a 75-day lagged moving average. These transformed values were then normalised using linear scaling into the range 0 to 1.

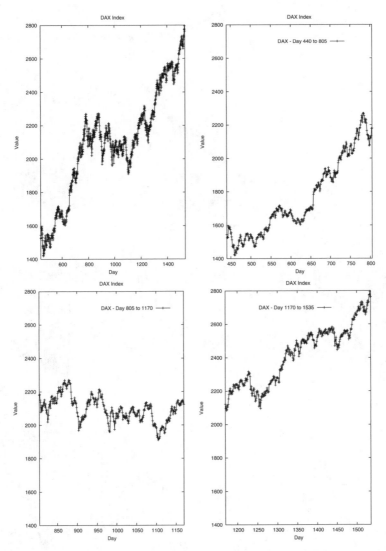

Fig. 13.1. A plot of the DAX over the entire data set (top left). Taken from this data set, we can see the training period (top right), and the two test periods (bottom left and right respectively)

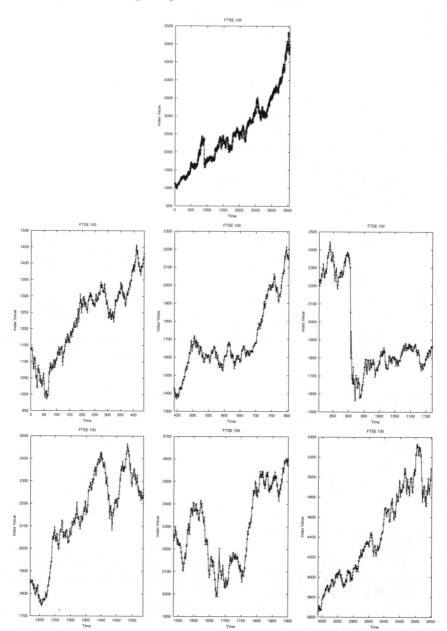

Fig. 13.2. A plot of the FTSE 100 over the entire dataset (top), over the training period (middle-left), and over each of the five out-of-sample test periods

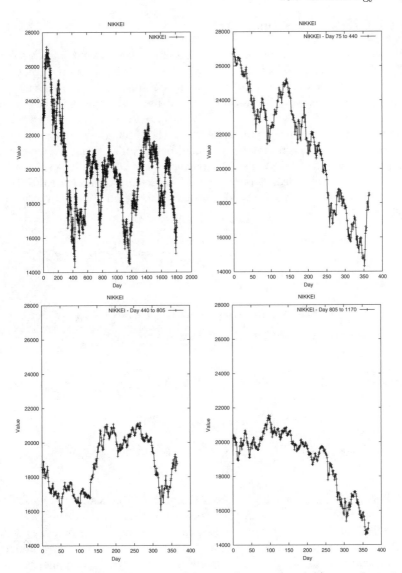

Fig. 13.3. A plot of the Nikkei over the entire data set (top left), over the training period (top right), and over the two test periods (bottom left and right respectively)

13.1.1 GE System Setup

When applying GE to a problem one must first decide what form a solution will take and design a grammar that will allow the construction of these solutions, providing a suitable set of functions, variables and syntactic rules. The grammar used to create the trading systems in this case study includes function definitions of moving average, momentum and trading range, permitting the inclusion of technical indicators using these functions in the generated trading systems. The grammar is outlined below. In this case we have explicitly stated the set of terminals (T), non-terminals (NT), and the start symbol (S), although in practice we only provide the set of production rules (P), the set of terminals and non-terminals and the start symbol being described implicitly within P.

```
N={<code>,<expr>,<fopbi>,<fopun>,<matbi>,<relbi>,<var>,<int>}
T={p,=,(,),f_and,f_or,f_not,+,-,*,>,<,>=,<=,scale,ma,day,1,2,3,4,5,10}
S=<code>
P={
<code> ::=  p = <expr> ;

<expr> ::= <fopbi> (<expr>, <expr>) | <fopun> (<expr>)
          | <expr><matbi><expr> | <expr><relbi><expr> | <var>

<fopbi> ::= f_and | f_or

<fopun> ::= f_not

<matbi> ::= + | - | *

<relbi> ::= > | < | >= | <=

<var> ::= <int> | day | ma(<int>,day) | momentum(<int>,day)
         | trb(<int>,day)

<int> ::= 1 | 2 | 3 | 4 | 5 | 10
}
```

In addition to the technical indicators the grammar also allows the use of the binary operators f_and, f_or, the standard arithmetic operators, the unary operator f_not, and the current days index value day. The operations f_and, f_or and f_not return the minimum, maximum of the arguments, and 1-the argument, respectively. The daily signals generated by the trading system are postprocessed using the following rule:

$$Buy = Value < 0.33$$
$$DoNothing = 0.33 >= Value < 0.66$$
$$Sell = 0.66 >= Value$$

Fitness Function

A key decision in constructing a trading system is to determine what fitness measure should be adopted. A simple fitness measure, such as the profitability of the system both in and out of sample or the excess return to a trading strategy as against a buy-and-hold strategy (where the fitness measure used in the evolutionary process is defined as the 'excess return'), is incomplete, as it fails to consider risk. The risk of a trading system can be estimated in a variety of ways. One method is to consider market risk, defined here as the risk of loss of funds due to an adverse market movement. A measure of this risk is provided by the maximum *drawdown* (the maximum cumulative loss) of the system during a training or test period. This measure of risk can be incorporated into the fitness function in a variety of formats including $\left(\frac{return}{maximum\ drawdown}\right)$ or $(return - x(maximum\ drawdown))$. Each fitness function will encourage the evolution of trading systems with good return to risk characteristics by discriminating against high-risk/high-reward trading rules. In the second version of the fitness function, x represents a *tuning parameter*. As the value of x is increased, the evolved trading systems become more conservative. In this case study, we use the second version of the fitness function, and set the value of x to one.

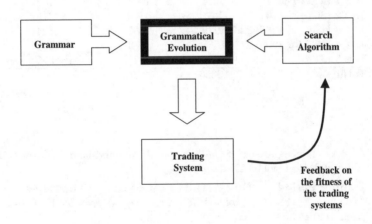

Fig. 13.4. GE system with plug-in inputs, the search algorithm (genetic algorithm) and the grammar

13.2 Results

Thirty runs were performed on each of the three datasets, each using a population size of 500 individuals over 100 generations. Roulette selection, steady-state replacement (two parents generate two children with the best child entering the next generation if it has better fitness than the worst member of the current population), bit mutation at a probability of 0.01, and a variable-length one-point crossover operator with a probability of 0.9 were adopted.

A comparison of the best individuals evolved for each dataset to the benchmark buy-and-hold strategy can be seen in Tables 13.1 to 13.3, for the FTSE, DAX and Nikkei datasets respectively. In the case of the FTSE and Nikkei datasets the evolved rules produce a superior performance to the benchmark strategy, while performance over the DAX dataset is not as strong. The poorer performance for the DAX market likely results from overfitting of the evolved rules to the training data. For each of the evolved trading rules the associated risk is less than that of the benchmark strategy as can be seen in the average daily investment figures reported.

Table 13.1. A comparison of the buy-and-hold benchmark to the best evolved individual for the FTSE dataset

Trading Period (Days)	Buy & Hold Profit (US$)	Best-of-run Profit (US$)	Best-of-run Avg. Daily Investment
Train (75 to 440)	3071	3156	3219
Test 1 (440 to 805)	5244	1607	1822
Test 2 (805 to 1170)	-1376	4710	3151
Test 3 (1170 to 1535)	1979	2387	6041
Test 4 (1535 to 1900)	1568	-173	3274
Test 5 (3196 to 3552)	3200	2221	3767
Total	13686	13908	

Table 13.2. A comparison of the benchmark buy-and-hold strategy to the best evolved individual on the DAX dataset

Trading Period (Days)	Buy & Hold Profit (US$)	Best-of-run Profit (US$)	Best-of-run Avg. Daily Investment
Train (440 to 805)	3835	3648	7548
Test 1 (805 to 1170)	-41	-1057	8178
Test 2 (1170 to 1535)	3016	469	8562
Total	6831	3060	

13.3 Discussion

The trading systems evolved by grammatical evolution demonstrate a performance superior to the benchmark buy-and-hold strategy on two of the three datasets. In addition, the risk involved with the adoption of the evolved trading rules is less than that of the benchmark. The risk of the benchmark

Table 13.3. A comparison of the benchmark buy-and-hold strategy to the best evolved individual on the Nikkei dataset

Trading Period (Days)	Buy & Hold Profit (US$)	Best-of-run Profit (US$)	Best-of-run Avg. Daily Investment
Train (75 to 440)	-6285	3227	9247
Test 1 (440 to 805)	59	-1115	7164
Test 2 (805 to 1170)	-3824	633	9192
Total	-10050	2745	

buy-and-hold portfolio exceeded that of the portfolio generated by the technical trading rules because the benchmark buy-and-hold portfolio maintains a fully invested position at all times in the market, whereas the portfolio generated by the evolved technical trading system averaged a capital investment of $3,546, $8,096 and $8,534 over the trading periods on the FTSE, DAX, and Nikkei datasets, respectively.

Refining the System

There is substantial potential to tweak the above model to improve its performance. The grammar only considers a small set of technical indicators. The incorporation of additional technical indicators and environmental information would further improve the performance of the evolved trading rules. A particular feature of the GE approach to creating trading systems is that it is easy to incorporate additional inputs and/or technical indicators into the system's grammar.

Another extension would be to incorporate a more complex model of learning (forgetting). Glassman [90] suggested that the 'fallibility of memory' (p. 88) may represent a useful adaptive device when faced with a dynamic environment. In the above illustration, all historic data observations are given equal weighting which implicitly assumes model stationarity. By a suitable modification of the fitness function, whereby more recent data observations are assigned a higher weighting in the model construction process, model development could be biased towards more recent data [184]. The weighting parameter could also be evolved as a component of the developed model. The next chapter demonstrates how grammatical evolution can be combined with a moving training window in order to develop an adaptive trading system.

In this case study a simple exit strategy, which automatically closed trading positions after ten days, was adopted. Therefore, in essence, the grammar was designed to evolve good trading position 'entry' strategies. The current model could easily be extended to also evolve the exit strategy by making appropriate modifications to the grammar. The importance of optimising the exit strategy should not be overlooked, and this issue is addressed in the intra-day trading case in chap. 15.

Portfolios of Trading Rules

A particular benefit of adopting a population-based approach to developing trading rules is that multiple rules are uncovered. Each of these rules has (hopefully) a reasonable chance of working well in the future. Rather than implementing a trading system which relies on a single rule, for example the best rule found, an obvious strategy is to diversify trading across a number of the better trading rules.

There are several methods of implementing this approach. The simplest is to allocate funds to each rule and trade them independently. Another approach is to implement a stacked or multi-stage model, which takes the trading signals produced by several rules, combines them, and produces a final trading signal. A variant on this approach is to create a series of 'families' of trading rules, using GE or an alternative methodology, where each family is trained using non-homogeneous inputs. The predictions from the best rule from each 'family' could then be used as inputs to a second-stage model which produces the final trading signal. Periodically the entire system could be retrained, and new trading rules created.

14

Adaptive Trading Using Grammatical Evolution

14.1 Introduction

Following on from Chap. 13, this case study illustrates the construction of an adaptive trading system using GE. Rather than employing a single fixed training period, the trading system continues to retrain as new data becomes available using a variant on the *moving window* approach. This permits the system to adapt to dynamic market conditions, while maintaining a memory of good solutions that worked well in past market environments. In contrast to the trading system developed in Chap. 13, the system can also adjust the size of the position it takes in the market depending on the strength of the trading signal which is produced.

14.2 Methodology

The trading system developed is based on moving averages. Simple extensions to the grammar, embedding other technical indicators, would permit the evolution of more complex sets of technical trading rules.

Grammar

```
<expressions> ::= <expressions> <op> <exp> | <exp>
<exp> ::= MA( <numExp> ) | <numExp>
<op> ::= - | + | * | /
<numExp> ::= <numbers> <op> <numbers> | <numbers>
<numbers> ::= <numbers><number> | <number>
<number> ::= 0 | 1 | 2 | 3 | 4 | 5 | 6 | 7 | 8 | 9
```

The trading rules evolved by GE using the above grammar (drawn from [57]), generate a signal that can result in one of three trading actions: Buy, Sell or Do Nothing. The <expressions> non-terminal produces the signal using combinations of the MA terminal which is a moving average function.

14.2.1 Moving Window

The adaptive training process commences in the same manner as in the last chapter. An initial training period is set aside on which the population of proto-trading rules is trained, with the aim of evolving a reasonably competent population of trading rules after a certain number of generations (G). The system then goes 'live', and begins to trade. The trading system takes the best performing rule from the initial training period, and uses this rule to trade for each of the following x days. After x days have elapsed, the *training window* moves forward in the time-series by x days, and the current population of trading rules is retrained over the new data window for a number of generations g, where $g < G$. This training process embeds both a memory and an adaptive capability in the trading system, as good past trading rules serve as a starting point for trading system adaptation.

The value of g relative to that of x determines the memory/adaptiveness balance in the trading system. A small value of g means that memory is emphasised over adaptation, as new data has relatively less chance to influence the trading rules. The value of g need not be fixed, and could itself be adapted over time. For example, in periods of rapid market change a trading system with a 'long memory' could be disadvantageous, whereas in stable periods a longer memory could well be advantageous. The length of the trading/re-training window (x) also impacts on the adaptiveness of the trading system. If x is large, the trading rules are altered less frequently, but each adaptive 'step' during retraining will tend to be larger.

In implementing the moving window training process in this case, the first 440 days data is used to create the initial population of trading rules. Data from days 1-75 is reserved to allow the evolved rules use moving averages of up to a maximum lag of 75 days. The trading rules are trained on the data for days 76-440, for 100 (G) generations. The trading rule which generates the best return over the training period is then used to trade 'live' (out of sample) for the next 5 days (x). The training window is then moved forward to include these 5 days, and the population of trading rules is adapted by retraining it for 2 or 10 (g) generations. Figure 14.1 provides a diagram of the training/live trading process.

14.2.2 Variable Position Trading

In the last chapter, the *entry strategy* for each trade was to invest a constant Dollar amount on the production of a buy or a sell signal. The relative strength of the buy or sell signal was not considered. In this case, the trading system adopts a more complex entry strategy, and a *variable size* investment is made, depending on the strength of the trading signal. The stronger the signal the greater the amount invested, subject to a maximum investment amount of $1,000 (arbitrary). The amount invested for each signal is:

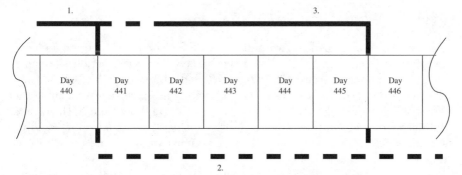

Fig. 14.1. The initial population of trading rules is evolved using data from days 1-440 (step 1 in the diagram). Next the best of these rules is used to trade live for five days (step 2). Finally, the training window is moved forward five days in the dataset, and the current population of rules is retrained for g generations (step 3)

$$Amount\ invested\ = \frac{Size\ of\ trading\ signal}{Maximum\ trading\ signal} * 1000 \qquad (14.1)$$

Signals received from a trading rule oscillate around a pivot of zero. Signals greater than zero constitute a buy signal, those less than zero constitute a sell signal. To allow the system to decide how much to invest on a given trade using the above rule, the maximum size of a trading signal must be determined, and we do this in an adaptive manner. Initially we set the maximum signal as being the size of the first buy signal generated by the system. If a signal is subsequently generated that is stronger than this, the maximum trading signal is reset to the new amount. If the sum to be invested is greater than the cash available, the model will invest the cash available less the costs of the transaction. Upon receipt of a sell signal all positions are closed.

14.2.3 Return Calculation

The total return generated by a trading system is calculated as a combination of its trading return (net of transactions costs) and a risk-free interest return on uninvested funds. Transaction costs are based on the cost structure used by online trading houses, where flat fees are incurred for the opening and closing of positions. A $10 fee is charged upon entry and exit of trades. The trading system is constrained from making very small trades (those less than $100) which would be uneconomic given the fixed transaction cost by including a rule which classes all uneconomic trades as a Do Nothing signal. The fitness measure adopted during training was the total return over the training period.

14.3 Results

The system was applied to two indices, the S&P 500 for the period January 1st 1991 to December 1st 1997, and the Nikkei 225 for the period December 10th 1992 to December 3rd 1997 (Figs. 14.2 and 14.3 display the S&P 500 and the Nikkei 225 over the relevant time periods). The experiments examine the effect of using two different values of g (2 and 10), and benchmark the results obtained from the adaptive trading system against those of a 'restart' trading system which retrains completely in each training window (and therefore has no memory).

A population size of 500 individuals was used with 100 generations of training for the initial period. A generational rank replacement strategy was used with 25% of the weakest performing members of the population being replaced with newly-generated individuals in each generation. Thirty runs were conducted for each of the experiments for each market, with a crossover rate of 0.9 and a mutation rate of 0.1 as in [57] and [59].

The rest of the results section is broken up into two parts. Section 14.3.1 reports the returns of each trading system over its training range versus the returns made by the index over the same period. Section 14.3.2 reports the returns made during out-of-sample trading.

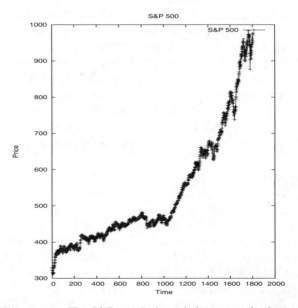

Fig. 14.2. The S&P 500 Index 1/1/1991 to 3/12/1997

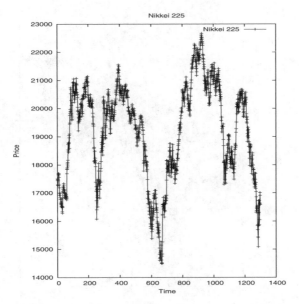

Fig. 14.3. The Nikkei 225 Index 10/12/1992 to 3/12/1997

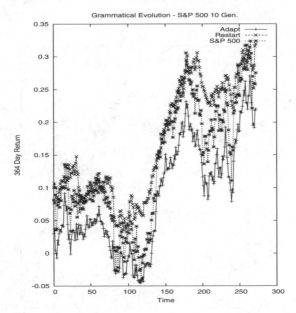

Fig. 14.4. Training performance on the S&P 500 for 10 generations of training at each window increment

14.3.1 Training Returns

As the live trading window consists of 5 days, the S&P 500 dataset produced 273 distinct retraining windows, and the Nikkei 225 dataset produced 171

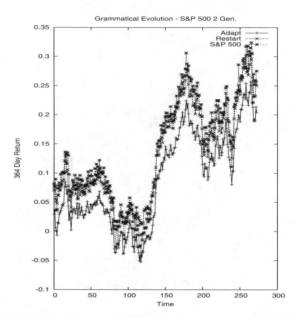

Fig. 14.5. Training performance on the S&P 500 for 2 generations of training at each window increment

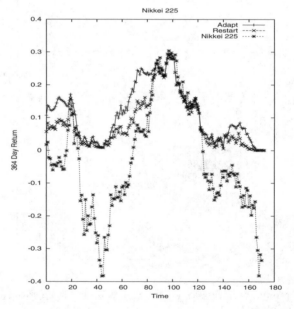

Fig. 14.6. Training performance on the Nikkei 225 for 10 generations of training at each window increment

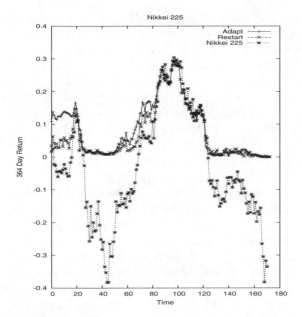

Fig. 14.7. Training performance on the Nikkei 225 for 2 generations of training at each window increment

distinct training windows. At the end of the final generation in each training period, the return of the best trading rule (based on its fitness over that training window) was determined. The values of the S&P 500 and Nikkei 225 indices at the start and end of each training window were also determined in order to calculate the return to a buy-and-hold investment strategy. Figures 14.4-14.7 compare the buy-and-hold returns from the S&P 500 and Nikkei 225 with those from the adaptive and the restart trading systems.

Examining these results using both a t-test and bootstrap t-test [70], we see a statistically significant difference between the returns generated by the adaptive and the restart training methodologies for the S&P 500. For the Nikkei 225 no statistical difference between the two training methods is found. This is partly due to the behaviour of the index towards the end of the dataset where it shows a loss of almost 40%. During this decline, the trading rules of both the adaptive and restart paradigms generally opted to remain outside the market and did not trade.

14.3.2 Out-of-Sample Returns

Tables 14.1 and 14.2 display the results and a breakdown of the out-of-sample trading for each of the sets of experiments conducted for the S&P 500 and Nikkei 225. For the S&P 500 the adaptive population experiments provided best returns of 63% and 64% for the 2 generation and 10 generation tests respectively, and average best returns of 48% and 44% respectively. The restart population experiments provide best returns of 59% and 41% and average best returns of 45% and 32% for the 2 generation and 10 generation tests respectively. When the out-of-sample results are compared with the training graphs in Sect. 14.3.1, in which the restart method tends to outperform the adaptive method, it appears that the restart populations have overfit the training data and are not generalising as well as when applied to the out-of-sample data. For the Nikkei 225 none of the developed trading systems made positive returns (the index itself made a return of −21%), but in all cases the best evolved trading systems comfortably outperformed a buy-and-hold strategy.

An interesting aspect in the comparison between the adaptive and restart paradigms is analysing the role that memory plays in the adaptive setup, a feature which is not present in the restart paradigm. For the S&P 500 experiments the 2 generation setup was seen to reuse trading rules from the previous trading window 26% of the time (on average), for the 10 generation setup this percentage of reuse grew to 32%. For the turbulent Nikkei 225 index the adaptive 2 generation setup reused rules 64% of the time, with the 10 generation setup reusing rules 52% of the time. What this highlights is that the adaptive method can (and is) taking advantage of previous learning. In contrast, the restart method must relearn good trading rules from scratch in each window increment.

Table 14.1. Profit, loss and trading analysis (out-of-sample) for each setup on the S&P 500

	2 Gen. Adapt	2 Gen. Restart	10 Gen. Adapt	10 Gen. Restart
Best (%)	63	59	64	41
Avg. Best (%)	48	45	44	32
Avg. No. Trades	125	110	174	209
Avg. Profitable Trades (%)	73	65	73	71

Table 14.2. Profit, loss and trading analysis (out-of-sample) for each setup on the Nikkei 225

	2 Gen. Adapt	2 Gen. Restart	10 Gen. Adapt	10 Gen. Restart
Best (%)	-0.15	-8	-10	-1.5
Avg. Best(%)	-16	-19	-25	-21
Avg. No. Trades	49	111	88	139
Avg. Profitable Trades (%)	35	47	30	45

14.4 Discussion

The case study illustrates the ability of grammatical evolution to evolve adaptive trading rules, which are shown to have an advantage over a static restart approach. It is worth noting that while the rules evolved for the Nikkei 225 did not provide profitable returns, they still beat the index's overall performance in three of the four experiments. In one instance the adaptive approach beat the index by 20.85% to be just 0.15% short of breaking-even, including trading costs.

There is much scope to enhance the approach presented here. For example, several parameters of the adaptive model in this study could be examined in more detail, including the effect of altering the number of generations evolved at each increment, and of altering the size of each window increment.

15

Intra-day Trading Using Grammatical Evolution

This case utilises high-frequency time-series data to construct an intra-day trading system for two stocks (Ford and IBM) using a grammatical evolution (GE) methodology. In order to illustrate the effect of different trade-exit strategies on the performance of a trading system, three different exit strategies are compared.

In exchange-traded markets such as equity markets, a *tick* represents a time-stamped record of a transaction on the exchange. This record includes the price and volume of each trade. Financial markets generate a huge quantity of tick data each day. An actively-traded share on a major exchange may trade multiple times per minute. A median stock in the Russell 3000 produces approximately 500,000 ticks per year, whereas a heavily traded share like Microsoft may generate 20 million ticks annually. Traders can see this data in real time and can use it in making trading decisions.

15.1 Background

High-frequency financial data, such as the data tackled in this case, is data which is sampled at small time intervals (at a high frequency) during the trading day. Therefore, high-frequency financial data can contain the price and volume history of a financial asset for every second of trading during the day. Substantial volumes of trading in financial markets are intra-day. In the case of foreign exchange markets, Dacorogna et al. [47] estimate that approximately 90% of all trading is accounted for by intra-day traders.

Most academic studies of the utility of technical analysis use low-frequency time-series, typically end-of-day price and volume information (a notable exception is Svangard et al. [209] who used market information sampled at a one-minute interval). There are approximately 250 trading days per year, so using a daily sampling frame results in a significant reduction in the density of data. For example, using end-of-day data for Microsoft implies that each (daily) data point represents on average 80,000 transactions.

Studies of technical analysis using daily price data provide limited insight into the possible utility of technical analysis for intra-day trading. If the intention is to construct a trading system that will hold positions for a short-horizon (intra-day) then using daily price data to construct the system will not suffice. Rather the system will need to be constructed using tick data, or a high-frequency time-series sampled from this data.

The use of high-frequency data allows the collection of large quantities of data for trading system construction over relatively short time periods. This provides scope for the rapid construction, testing and live-trading of a system. Over short time spans the fundamental assumption of any trading system, that the present resembles the past, is more likely to be tenable. Real-time tick-by-tick data feeds are easily and inexpensively available from several data vendors.

15.2 Methodology

The trading systems in this case study were developed using high-frequency price data for Ford and IBM drawn from the period 2/1/02 to 15/4/02, and were tested out of sample using data from 16/4/02 to 2/7/02. Data is sampled at five-minute intervals from these periods, producing 9,828 individual data points for each stock. Each data point included the opening and closing prices, the high and low price and volume, for each five-minute interval. Figure 15.1 provides a *high-low close (HLC) chart* for Ford, for the first 50 minutes of trading on 2/1/2002 (split into ten 5 minute trading intervals). The top and bottom of each bar represents the high and low price for each interval, and the horizontal tick represents the closing price. The average of the open and closing price of each interval is used as the input data for the trading system. The price data is normalised into the range [0,1].

Characteristics of Tick-by-Tick Data

It is well known that tick-by-tick data exhibits intra-day seasonalities. In the case of equity and bond markets which are open for fixed trading hours, activity measures display a distorted U shape over the trading day [8]. Typically, intra-tick duration (the time between one tick and the next) is lowest in the opening minutes of daily trading. As an example of the volume of trading which can occur at the start of a trading day, a total of 3,352 trades were recorded for Microsoft in the first 180 seconds of trading (9.30-9.32am) on 20/5/02. In contrast, the average number of ticks per 180 seconds of trading over the entire year was approximately 620. Intra-tick duration is highest during lunch hour and decreases again towards the end of the trading day. Figure 15.2 provides a graph of the average trading volume of Ford and IBM for each 5 minute interval during the trading day, for the period January-July 2002. A practical consequence is that sampled data from a tick-by-tick

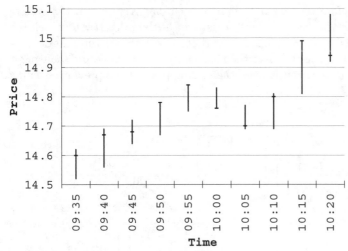

Fig. 15.1. High-low close chart of Ford, for first ten, five-minute intervals of trading, on 1st February 2002 (x-axis is denominated in dollars)

series may not contain homogeneous information. Consider a sampled price at 9.40 am versus a sampled price at 1.20 pm. It is likely that the price at 9.40 am is representative of a number of trades that took place in the active early-morning session. The later sampled price may reflect far fewer trades.

In addition to the predictable pattern of market trading intra-day, many markets also display a tendency to reach their high and low points during either early or late trading during the day. Prices tend to be most volatile in early and late trading, due to the queuing of pre-opening trades in the morning, and the activities of traders who do/do not want to carry inventory of a stock overnight in the case of late trading.

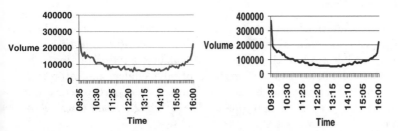

Fig. 15.2. Average intra-day trading volume for Ford (left) and IBM (right) at 5 minute intervals, Jan-July 2002

In developing our trading system, to overcome these periods of volatility we do not trade in either the first or last half hour each day. All open positions are closed out before the last half hour of each day, resulting in the system not holding any positions overnight. This reduces the price risk to which the system is exposed, as new information that is brought to the market pre-opening the next day, could affect the next morning's opening price causing it to gap upwards or downwards from the closing price of the previous evening (Fig. 15.3).

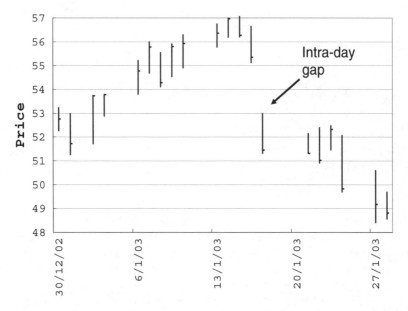

Fig. 15.3. Example of intra-day gap for Microsoft (16-17th January 2003) (*x*-axis is denominated in dollars)

15.2.1 Trading System

After the initial half-hour period each day, the trading system considers whether or not to trade at the end of every 10 minute interval during each day. At each possible trading time the system calculates its prediction. The prediction is calculated by evaluating the evolved technical indicator rule. The trading rule returns a value in the range 0 to 1, and postprocesses this value, as in the last case study, into one of three trading signals *Buy*, *Sell* or *Do-Nothing*.

If the prediction is to go long the system will buy $1,000 of stock, if it is to go short the system will sell $1,000 of stock. When the trade is closed out a profit or loss is evaluated, and a cumulative total of the profits or losses of the trading rule is maintained. The maximum amount that the system can have invested at any one time is $10,000. If the total trading capital is invested at any time, no further positions are open until preexisting positions are closed.

Exit Strategies

Once a trading position is opened, a variety of *exit strategies* could be employed to decide when this position is to be closed out. In order to examine the significance of the choice of exit strategy on the results obtained by a trading system, this study evolves trading systems which use three different exit strategies:

- standard close,
- extended close, and
- stop-loss, take-profit close.

The standard and extended close strategies are examined for both stocks, and the stop-loss, take-profit close strategy is examined for Ford. In the *standard close*, the evolved systems automatically close out all trading positions 30 minutes after they are opened. In the *extended close*, the system rechecks after 30 minutes whether the prediction is unchanged from the initial prediction, and if it is the trade is extended for a further 30 minutes. In the *stop-loss, take-profit close*, the position is initially held for 30 minutes, and thereafter, if the position generates a loss of 0.1% it is closed immediately, and profit is automatically taken on any position which makes a profit of 0.8% by closing the position once the take-profit trigger is hit. If the position is still open 30 minutes from the end of the trading day it is closed out.

15.2.2 GE System Setup

The grammar adopted in the GE system is defined as follows:

```
<code> ::=  p = <expr> ;

<expr> ::= <fopbi>(<expr>,<expr>) | <fopun>(<expr>)
         | <expr><matbi><expr> | <expr><relbi><expr>
         | <var>

<fopbi> ::= f_and | f_or

<fopun> ::= f_not

<matbi> ::= + | - | *

<relbi> ::= > | < | >= | <=

<var> ::= <int> | day | ma(<int>,day)
        | momentum(<int>,day) | trb(<int>,day)

<int> ::= 1 | 2 | 3 | 4 | 5 | 10
```

In addition to the technical indicators the grammar also allows the use of the binary operators f_and, f_or, the standard arithmetic operators, the unary operator f_not, and the current days index value day. The operations f_and, f_or, and f_not return the minimum and maximum of the arguments, and 1-the argument, respectively.

The GA algorithm in the GE system uses roulette selection, a steady-state replacement mechanism such that two parents produce two children, the best of which replaces the worst individual in the current population, if the child has greater fitness. The standard genetic operators of bit mutation (probability of 0.01) and crossover (probability of 0.9) are adopted. At the end of each run the best individual is stored, along with that individual's trading rule, fitness and drawdown.

15.3 Results

The results from our experiments are now provided for both the training period (Table 15.1) and the test period (Table 15.2). Neither stock displayed a distinct monotonic price trend during the train or test periods (see Fig. 15.3 for a graph of Ford's share price over the train/test period).

Table 15.1. Trading profit (maximum drawdown) in $ during the training period

	Ford	IBM
Standard Close	1,280.73 (1.02)	1,221.67 (12.08)
Extended Close	2,436.67 (29.52)	2,431.51 (68.29)
Stop-Loss, Take-Profit Close	1,965.07 (1.01)	n/a
Buy-and-Hold	96.92	-2,890.28

Table 15.2. Trading profit (maximum drawdown) in $ during the test (out-of-sample) period

	Ford	IBM
Standard Close	823.86 (14.47)	892.50 (20.42)
Extended Close	1,257.71 (61.96)	1,761.33 (2.21)
Stop-Loss, Take-Profit Close	1,291.10 (14.47)	n/a
Buy-and-Hold	286.55	-1,905.94

In both the training and test periods, the extended close exit strategy notably outperforms the standard close strategy, without exhibiting clearly higher drawdowns. This result highlights the impact that the choice of exit strategy can have on the results produced by a trading system. In each case, the trading systems were developed on the same data, using the same GE algorithm, with only the exit strategy differing. In considering the utility of the stop-loss, take-profit exit strategy (which has only been examined for Ford), it is noted that it outperforms the standard close exit mechanism in both the training and

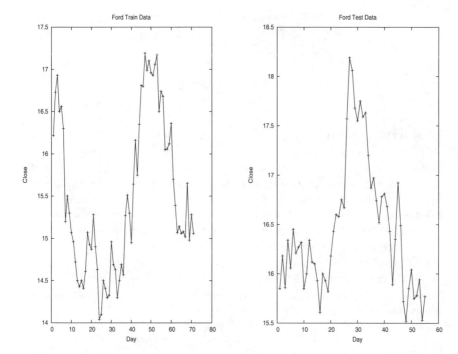

Fig. 15.4. Ford price during train and test periods

test periods. It does not clearly dominate the extended close exit strategy, underperforming in the training period, and outperforming in the test period. Table 15.3 provides information on the percentage of profitable trades under each exit strategy for the test period. Under the standard close approximately 55% of trades are winning trades, but this percentage drops under the other two strategies, notably under the stop-loss, take-profit strategy. Despite the lower percentage success under the latter two strategies, they substantially outperform the standard close strategy in terms of dollar profit generated. This illustrates the danger in using a simple metric such as percentage of successful trades as a fitness measure in evolving trading systems. In the case of the stop-loss, take-profit strategy, the low percentage of profitable trades is explained by the tight take-loss criterion, where positions losing more than 0.1% were closed out. This has the effect of closing out positions which have incurred a small loss, increasing the percentage of trades closed at a loss, but simultaneously reducing the price risk of the trading system. Although not undertaken in this example, the trigger points for the stop-loss and take-profit could themselves be evolved as part of the trading system.

Table 15.3. Percentage of profitable trades in test period

	Ford	IBM
Standard Close	55	54
Extended Close	45	55
Stop-Loss, Take-Profit Close	32	n/a

15.4 Discussion

In this case, GE was applied for the purposes of evolving intra-day trading systems for equity markets. GE was shown to evolve profitable trading rules for both training and test periods, and the evolved trading rules did not exhibit large drawdowns. The significance of the choice of exit strategy for the profitability of the evolved trading systems was also illustrated.

A large number of extensions of the methodology are possible. In evolving the trading systems, we ignored trading costs and slippage, and did not explicitly incorporate a risk penalty into the fitness function. There is also scope to develop more sophisticated money-management strategies, as seen in Chap. 14.

16

Automatic Generation of Foreign Exchange Trading Rules

The prediction of foreign exchange rates is a difficult task. Many interconnected political and macroeconomic factors (including inflation rates, interest rates, money supply and balance of payments) impact on the fundamental value of a currency, and markets will respond as expectations as to future values of these items alter. Foreign exchange rates are a function of current levels of demand and supply, and technical analysis can also be applied to these markets. This case illustrates the application of grammatical evolution to uncover a series of useful technical trading rules which can be used to trade spot foreign exchange markets (transactions in which two currencies are immediately exchanged).

International foreign exchange markets are dominated by a small number of heavily traded currency pairings. Approximately 89% of all foreign currency transactions involve the US Dollar [20]. Prior to the introduction of the Euro, the most-traded currency pairings were the US Dollar/DM, US Dollar/Yen and US Dollar/Sterling, accounting for 20%, 18% and 8% of total daily average volume in currency markets during 1998 [19]. This case develops trading systems for each of these currency pairs using daily US-DM, US-Stg and US-Yen exchange rates for the period 23/10/92 to 13/10/97.

16.1 Background

Foreign exchange markets are the most active of all financial markets with average *daily* trading volumes in traditional (non-electronic broker) foreign exchange markets estimated at $1.9 trillion in 2004 [20]. Average daily trading on spot markets exceeds $621 billion. Unlike equity markets, foreign exchange markets are not exchange-traded, rather trades are bilateral agreements between individual parties who are not required to register these trades with any central agency. Also, unlike equity markets, there can never be a bear market for foreign exchange. As currencies are valued relative to one another, as one currency goes down another must go up.

Although the precise scale of speculative trading on spot markets is unknown, the bulk of transactions are not represented by world trade in goods and services [85]. Only about 17% of the trading is driven by non-dealer/financial institution trading [20]. Therefore, it is plausible to assume that the scale of speculative trading is substantial. Speculative foreign currency trading implies the existence of predictive models in the mind of investors. These models could integrate many different forms of information. This case study focuses on a subset of this information, the information contained in the historical time-series of exchange rates.

Use of Technical Analysis in the Forex Market

In a study conducted on behalf of the Bank of England [211], it was found that approximately 90% of financial institutions dealing in foreign exchange in London (the largest foreign exchange market in the world, with 31% of total trading volume [20]) placed some weight on information obtained from technical analysis in making trading decisions. Lui and Mole [145] reported similar findings among foreign exchange dealers in Hong Kong. Additional support for a technical analysis argument is found in Osler [179], where evidence of clustering of currency stop-loss and take-profit orders is noted. Clusters of take-profit orders were found at exchange rates characterised by *round numbers* (ending in 00) with stop-loss orders clustering just beyond round numbers. The effect of clustering of these orders is that trends in exchange rates will tend to reverse at predictable support and resistance levels, and trends tend to accelerate once these levels are breached.

16.2 Methodology

The methodology adopted is based on that in [30]. Daily closing exchange rate data is drawn from the London market for the period 23/10/92 to 13/10/97. The training data set was comprised of the first 799 trading days of the data set. The developed models are tested using an out-of-sample dataset of 548 trading days. The out-of-sample data is divided into two hold-out samples (274*2) to allow comparison of the hold-out results across different market conditions. As in the last case study, the rules evolved by GE are used to generate one of three trading signals for each day of the training and test periods. The possible signals are *Buy*, *Sell* or *Do-Nothing*.

Trading System

If a buy signal is indicated on a given day, a fixed sum investment is made in the foreign currency using borrowed funds. This position is automatically closed at the end of a five-day trading period. On the production of a sell signal, a fixed sum of the foreign currency is sold, and again this position is closed out after a five-day period.

Consider a daily observed exchange rate series S_t where $t = 1, 2, \ldots n$. The total return (measured as a percentage) generated by the developed trading system is a combination of its trading return net of trading costs and an interest differential. We define the five-day (%) trading return (before trading costs) for a long position as $(\frac{S_t}{S_{t+5}} - 1)$ and as $-1[1 - (\frac{S_{t+5}}{S_t})]$ for a short position.

In this trading strategy, the home currency is the US dollar. When a long position is taken, the foreign currency is purchased and the return equals the trading return plus the interest earned in the foreign currency, less the cost of the home funds borrowed to invest in the foreign currency. When a short signal is produced by the trading system, the foreign currency is borrowed, at a cost of the foreign interest rate, and is invested in US dollars for the trading period. The dollar interest rate is earned on these funds. A trading cost c of 0.025% and a slippage allowance of 0.01% in each (single) direction is included [139].

Incorporating Risk into the Fitness Function

To encourage the evolution of a trading system with good risk to return characteristics, we penalise trading systems leading to volatile patterns of returns. The risk of catastrophic loss is reduced by incorporating a measure of the system's drawdown (the maximum accumulated loss produced by the trading system during the training period) into the fitness function. Therefore, the function is defined as:

$$Fitness = Return - (Maximum\ drawdown) \qquad (16.1)$$

Many variants on this rule exist, and the characteristics of the final trading system will critically depend on the choice of fitness function. An alternative function could be:

$$Fitness = \frac{Return}{Maximum\ drawdown} * Win\ ratio \qquad (16.2)$$

In this case, the development of the trading system is biased towards winning trades, potentially leading to a system which trades infrequently but which tends to win the trades it makes.

GE System Setup

The grammar adopted in this study is as follows:

```
<code> ::=  p = <expr> ;

<expr> ::=  <fopbi> ( <expr> , <expr> )
         |  <fopun> ( <expr> )
```

```
          | <expr> <matbi> <expr>
          | <expr> <relbi> <expr>
          | <var>

<fopbi> ::= f_and
          | f_or

<fopun> ::= f_not

<matbi> ::= + | - | *

<relbi> ::= > | < | >= |<=

<var> ::= <int>
        | ma( <int> , day )
        | day
        | momentum( <int> , day )
        | trb( <int> , day )

<int> ::= 1 | 2 | 3 | 4 | 5 | 10
```

The technical indicators adopted in this case are the moving average, momentum and trading range breakout where the relevant periods and combination of indicators used are to be determined by evolution. Apart from the raw components of the technical indicators adopted (average, moving average, lag, each of which takes a real-valued argument representing the chosen time-window) the BNF grammar also allows the use of standard arithmetic operators (+, -, * and inequality operators) and the binary operators f_and, f_or, and the unary operator f_not. The current day's price is also provided as a model input.

Roulette selection with a steady-state replacement mechanism is used, such that two parents produce two children, the best of which replaces the worst individual in the current population if the child has greater fitness. The standard genetic operators of bit mutation (probability of 0.01) and crossover (probability of 0.9) are adopted as is a duplication operator (probability of 0.01) that duplicates a random codon and inserts this into the penultimate codon position on the genome. A series of technical indicators, are predefined as are a series of mathematical operators, each of which is made available for inclusion in candidate solutions with their incorporation into the grammar input into GE, as described above.

16.3 Results

A plot of the exchange rates of the DM, Stg and the Yen against the US$ over the training and test periods can be seen in Fig. 16.1. Examining the three exchange rates over the training and test periods does not reveal obvious monotonic trends. Therefore, the three exchange rates should provide a good test-bed for the GE models.

Thirty independent runs of the GE algorithm were conducted, each with a population size of 500 individuals and an evolutionary period of 50 generations. To assess the quality of the results obtained they are compared against a buy-and-hold benchmark. The buy-and-hold benchmark is determined by

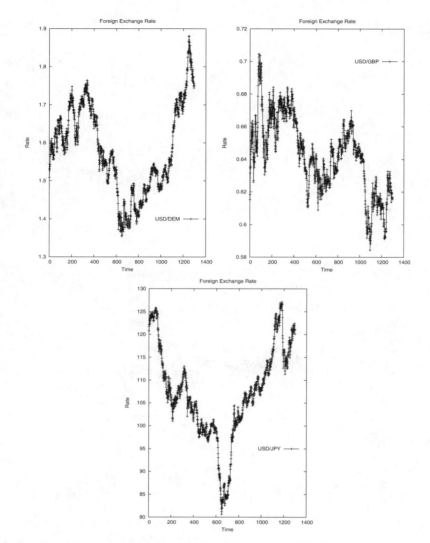

Fig. 16.1. Exchange rates for US$-DM, US$-Stg and US$-Yen over the training and test periods

comparing the return obtained from investing $5,000 in the foreign currency, net of the interest rate differential, during each training and out-of-sample period. A plot of the best individuals and mean of the best individuals' fitness at each generation over the 30 runs can be seen in Fig. 16.2.

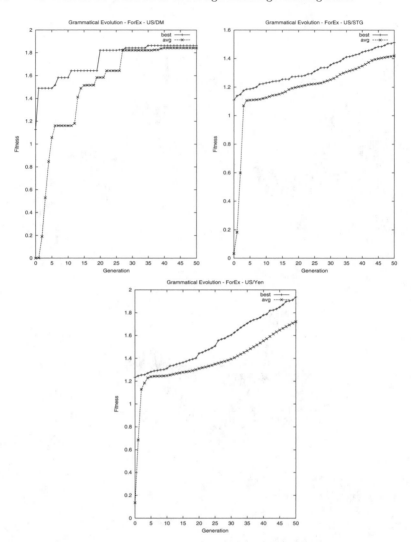

Fig. 16.2. In-sample fitness of overall best individual and mean of best individuals over the 30 runs, plotted for each currency pair

16.3.1 US-STG

Table 16.1 provides a comparison of the performance (a percentage return metric, $0.01 = 1\%$) of the best evolved trading rules against that of the benchmark investment strategy. The evolved rules outperform the benchmark strategy on the training and both test data sets.

Table 16.1. Results for the best (mean) evolved trading rules over the 30 runs compared to the benchmark buy-and-hold strategy on the US-STG dataset

Trading Period	Evolved Rule set	Buy-and-Hold
Training	1.75642 (1.5123)	-0.055855
Test 1	0.343709 (0.26103)	-0.094723
Test 2	0.500649 (0.47824)	-0.041456

16.3.2 US-Yen

Table 16.2 provides a comparison of the performance of the best evolved trading rules against that of the benchmark investment strategy. As for the US-STG exchange rate, the rules evolved by GE outperform the benchmark strategy on each of the datasets.

Table 16.2. Results for the best (mean) evolved trading rules over the 30 runs compared to the benchmark buy-and-hold strategy on the US-Yen dataset

Trading Period	Evolved Rule set	Buy-and-Hold
Training	2.3803 (1.9374)	-0.080146
Test 1	0.342609 (0.066474)	0.133476
Test 2	0.195778 (0.32521)	0.090996

16.3.3 US-DM

Table 16.3 provides a comparison of the performance (a percentage return metric, 0.01 = 1%) of the best evolved trading rules against that of the benchmark investment strategy. In this case the evolved rules outperform the benchmark strategy on the training and second test data sets, and manage to generate positive returns on the other test data set.

Table 16.3. Results for the best (mean) evolved trading rules over the 30 runs compared to the benchmark buy-and-hold strategy on the US-DM dataset

Trading Period	Evolved Rule set	Buy-and-Hold
Training	2.02357 (1.865)	-0.112199
Test 1	0.017998 (0.017998)	0.068231
Test 2	0.302671 (0.302671)	0.153231

16.4 Discussion

A total of 799 days of data were used to train the models, which were then tested on a further 548 days of out-of-sample data. Despite the lengthy out-of-sample period, the evolved trading rules generated positive returns on all the hold-out samples, after allowance for trading, slippage and net interest costs. In five of the six hold-out periods, the best evolved rule outperforms the benchmark buy-and-hold strategy. The mean (over all 30 runs) best evolved rule outperforms the benchmark in four of the six hold-out periods. It is also notable that the out-of-sample results appear robust over out-of-sample periods, and do not indicate that the performance of the trading system is declining.

Corporate Failure Prediction Using
Grammatical Evolution

This case illustrates how grammatical evolution can be used to uncover a series of useful rules which can assist in the prediction of corporate failure, using information drawn from financial statements.

Corporate failure is a natural component of the market economy, facilitating the recycling of financial, human and physical resources into more productive organisations [67, 192]. Nonetheless, corporate bankruptcy can impose significant private costs on many parties including shareholders, providers of debt finance, employees, suppliers, customers, managers and auditors. All of these stakeholders have an interest in being able to identify whether a firm is on a trajectory which is tending towards corporate failure. Early identification of such a trajectory could facilitate successful intervention, to avert disaster. Corporate failure can arise for many reasons. It may result from a single catastrophic event or it may be the terminal point of a lengthy process of decline. Under the second perspective, corporate failure is a process which is rooted in management defects, resulting in poor decisions, leading to financial deterioration and finally corporate collapse [4, 104, 151].

Most attempts to predict corporate failure implicitly assume that management decisions critically impact on firm performance [11]. Although management decisions are not directly observable, their consequent effect on the financial health of the firm can be observed through their impact on the firm's financial ratios. Typically when constructing corporate failure prediction models, explanatory variables are drawn from the financial statements of the firm, from financial markets, general macroeconomic variables, and non-financial, firm-specific information). In this case study, attention is restricted to information drawn from financial statements.

There are a number of reasons to suppose that GE can prove fruitful in the prediction of corporate failure. The problem domain is characterised by a lack of a strong theoretical framework, with many plausible, competing explanatory variables. The selection of quality explanatory variables and model form represents a high-dimensional combinatorial problem, giving rise to potential for GE. Another useful feature of a GE approach is that it produces human-readable rules that have the potential to enhance understanding of the problem domain.

17.1 Background

Research into the prediction of corporate failure has a long history [78, 111, 199]. Early statistical studies such as Beaver [17] adopted a univariate methodology, identifying which accounting ratios had greatest classification accuracy in separating failing and non-failing firms. Although this approach did demonstrate classification power, it suffers from the shortcoming that a single weak financial ratio may be offset (or exacerbated) by the strength (or weakness) of other financial ratios. Altman [3] addressed this issue by developing a multivariate LDA model and this was found to improve the classification accuracy of the developed models. Altman's discriminant function was:

$$Z = 0.012X_1 + 0.014X_2 + 0.033X_3 + 0.006X_4 + 0.999X_5 \qquad (17.1)$$

where:
X_1 = working capital to total assets
X_2 = retained earnings to total assets
X_3 = earnings before interest and taxes to total assets
X_4 = market value of equity to book value of total debt
X_5 = sales to total assets

Other statistical methodologies which have been applied include logit and probit regression models [89, 165, 227], neural networks [193, 195, 219] and genetic algorithms [135, 217], and NN-GA hybrids [28].

17.1.1 Definition of Corporate Failure

No unique definition of corporate failure exists [4]. Possible definitions range from failure to earn an economic rate of return on invested capital, to legal bankruptcy followed by liquidation of the firm's assets. Typically, financial failure occurs when a firm incurs liabilities which cannot be repaid from liquid financial resources. However, this may represent the end of a period of financial decline, characterised by a series of losses and reducing liquidity. Altman [3] suggested that the primary cause of corporate failure is the failure of management to recognise the symptoms of decline in time to take remedial action. Any attempt to uniquely define corporate failure is likely to prove problematic. While few publicly quoted companies fail in any given year (Morris [158] suggests that the rate is below 2% in the UK, Zmijewski [227] reports that this rate is less than 0.75% in the US), poorer performers are liable to acquisition by more successful firms. Thus, two firms may show a similar financial trajectory towards failure, but one firm may be acquired and 'turned around' whilst the other may fail.

The definition of corporate failure adopted in this case study is the court filing of a firm under Chapter 7 or Chapter 11 of the US bankruptcy code. The selection of this definition provides an objective benchmark as the occurrence (and timing) of either of these events can be determined through examination of regulatory filings. Chapter 7 of the US bankruptcy code covers corporate liquidations, and Chapter 11 covers corporate reorganisations which usually follow a period of financial distress. Under Chapter 11, management is required to file a reorganisation plan in bankruptcy court and seek approval for this plan. On filing the bankruptcy petition, the firm becomes a *debtor in possession*. Management continues to run the day-to-day business operations but all significant business decisions must be approved by a bankruptcy court. Even if creditors or stockholders vote to reject the proposed reorganisation plan, the court may still confirm the plan (known as a *cramdown*) if it finds that the plan treats creditors and stockholders fairly. In most cases, Chapter 11 reorganisations involve significant financial losses for both shareholders [188] and creditors [77] of the distressed firm. Moulton and Thomas [159], in a study of the outcomes of Chapter 11 filings, found that there were few successful reorganisations, despite a perception that some management teams were using Chapter 11 filings as a deliberate strategy for dealing with certain firm specific events such as onerous labor contracts or product-liability claims.

17.1.2 Explanatory Variables

Five groupings of explanatory variables, drawn from financial statements, are given prominence in prior literature: liquidity, debt, profitability, activity/efficiency, and size [6]. Liquidity refers to the availability of cash resources to meet short-term cash requirements. Debt measures focus on the relative mix of funding provided by shareholders and lenders. Profitability considers the rate of return generated by a firm in relation to its size, as measured by sales revenue and/or asset base. Activity measures consider the operational efficiency of the firm in collecting cash, managing stocks and controlling its production or service process. Firm size provides information on both the sales revenue and asset scale of the firm and acts as a proxy metric on firm history [140]. The groupings of potential explanatory variables can be represented by a wide range of individual financial ratios, each with slightly differing information content. The groupings are interconnected, as weak (or strong) financial performance in one area will impact on another. For example, a firm with a high level of debt may have lower profitability due to high interest costs.

Whatever modelling methodology is applied in order to predict corporate distress, the initial problem is to select a quality set of model inputs from a wide array of possible financial ratios and then to combine these ratios using suitable weightings in order to construct a high-quality classifier. Given the large search space, of both inputs and model form, an evolutionary algorithm such as GE has particular promise.

17.2 Methodology

A similar methodology to [32] is adopted. A sample of 178 (89 failed and 89 non-failed) publicly quoted US firms was drawn from the period 1991 to 2000 in order to train and test the model. Only firms with sales exceeding $1M, which had existed for at least three years prior to entry into Chapter 7 or Chapter 11 and which were outside the financial sector were considered for inclusion in the sample.[1] Twenty-two potential explanatory variables, were collected for each firm for the three years prior to entry into Chapter 7 or Chapter 11. The date of entry into Chapter 7 or Chapter 11 was determined by examining regulatory filings for each firm. For every failing firm, a matched non-failing firm was selected. Failed and non-failed firms were matched both by industry sector and size (sales revenue three years prior to failure). The set of 178 matched firms was randomly divided into model building (128 firms) and out-of-sample (50 firms) datasets. The dependent variable is binary (0,1), representing either a non-failed or a failed firm.

Choosing the Explanatory Variables

The choice of explanatory variables is hindered by the lack of a clear theoretical framework which explains corporate failure [11, 215, 219]. Most empirical work on corporate failure adopts an ad hoc approach to variable selection. Prior to the selection of the potential explanatory variables for inclusion in this study, a total of ten previous studies were examined [3, 7, 12, 17, 48, 118, 157, 165, 193, 208]. These studies employed a total of 58 distinct ratios. A subset of 22 of the most commonly used financial ratios was selected for this study. The selected ratios were:

  i. EBIT/Sales
 ii. EBITDA/Sales
iii. EBIT/Total Assets
 iv. Gross Profit/Sales
  v. Net Income/Total Assets
 vi. Net Income/Sales
vii. Return on Assets
viii. Return on Equity
 ix. Return on Investment
  x. Cash/Sales
 xi. Sales/Total Assets
xii. Inventory/Cost of Goods Sold

---

[1]Financial firms were excluded on grounds of lack of comparability of their financial ratios with other firms in the sample.

17.2.1 GE System Setup

The classification accuracy of the developed models is assessed based on the overall classification accuracy arising in both the model-building and out-of-sample datasets (only classification accuracy on the model-building data was used in the construction of the models). For simplicity, the cost of each type of classification error is assumed to be symmetric. The fitness function could be easily altered to bias the model development process to minimise a specific type of classification error if required.

In general, the construction of classifier systems consists of two components, the determination of a *valuation rule* which is applied to each observation and the determination of a *cut-off* value. The grammar adopted in this study is as follows, and its output is interpreted (postprocessed) using a fixed 0.5 cut-off value to produce a classification (values equal to or above 0.5 are interpreted as a failing company, values below 0.5 are interpreted as a non-failing company).

```
<lc> ::= output = <expr> ;

<expr> :: ( <expr> ) + ( <expr> )
       | <coeff> * <var>

<var> ::= var1[index]  | var2[index]  | var3[index]
       | var4[index]  | var5[index]  | var6[index]
       | var7[index]  | var8[index]  | var9[index]
       | var10[index] | var11[index] | var12[index]
       | var13[index] | var14[index] | var15[index]
       | var16[index] | var17[index] | var18[index]
       | var19[index] | var20[index] | var21[index]
       | var22[index]

<coeff> ::=  ( <coeff> ) <op> ( <coeff> )
         | <float>

<op> ::= + | - | *

<float> ::= 20 | -20 | 10 | -10 | 5 | -5 | 4 | -4
         | 3 | -3 | 2 | -2 | 1 | -1 | .1 | -.1
```

The above grammar generates classifiers of the form:

```
output = (<some expression>*varX) + (<some expression>*varY) + ...
```

However, any combination and number of the 22 explanatory variables can be exploited by an evolved classifier. Due to the way the grammar is defined, the generated rules have a linear form. Hence this case also provides an illustration of how GE can be used to generate a linear model. The grammar definition could be easily altered to allow the construction of non-linear models by including non-linear functions in the grammar.

17.2.2 LDA Method

The results obtained from the GE classifier are benchmarked against results arising from a Linear discriminant analysis (LDA) methodology. LDA derives a linear combination of characteristics (variables) which best discriminates between a series of predefined classes. The discriminant function may be expressed as follows for the two-class case: $Z = V_1X_1 + V_2X_2 + ... + V_nX_n$, where $V_1, V_2, ..., V_n$ are the discriminant coefficients and $X_1, X_2, ..., X_n$ are the independent variables. The function transforms a data vector into a single discriminant value which is then used to classify the observation.

The LDA model rests on a number of assumptions. To ensure the technique can produce an optimal classification rule, the data for each classification group must be drawn from a multivariate normal population and the covariance matrices must be homogeneous [147]. These assumptions do not generally hold for financial ratio data. Despite this, the technique has produced good results in the prediction of corporate distress using such data.

17.3 Results

Three series of models were constructed using explanatory variables drawn from one, two and three years (T1, T2 and T3) prior to failure. For each set of models, 30 runs were conducted using population sizes of 500, running for 100 generations, adopting one-point crossover at a probability of 0.9, and bit mutation at 0.01, along with roulette selection and a steady-state replacement strategy. A plot of the mean best and mean average fitness at each generation over the 30 runs can be seen in Fig. 17.1.

The classification results of the evolved models show promise. Despite drawing a sample from a variety of industrial sectors, the models demonstrate a high classification accuracy in and out-of-sample, which degrades gracefully rather than suddenly in the third year prior to failure. The best individuals evolved for each period are reported in Table 17.1. Calculation of Press's Q statistic [103] for each of these models rejects a null hypothesis, at the 5% level, that the out-of-sample classification accuracies are not significantly better than chance. In sample it can be seen that the classification performance of the models degrades as we move out each year. It is interesting to note that

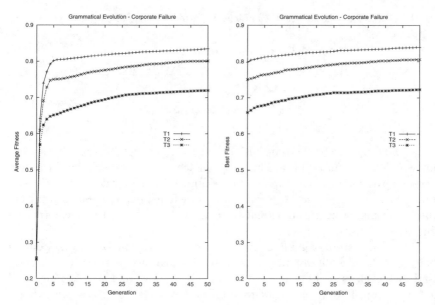

Fig. 17.1. In-sample fitness of overall best individual and mean of best individuals over the 30 runs

out of sample there is no performance difference between the evolved models in periods T1 and T2, both giving 80% correct classifications.

17.3.1 Form of the Evolved Classifiers

The best classifiers evolved for each period are:
One Year Prior to Failure:

```
Output = -3*Financial leverage -5*Return on Assets
+3*Inventory/Working Capital-20*Retained Earnings/Total Assets
+4*Total Liabilities/Total Assets
```

Two Years Prior to Failure:

```
Output = -2*Return on Assets+10*Sales/Total Assets-10*Fixed
Assets/Total Assets-2*varEBIT/Interest
```

Three Years Prior to Failure:

```
Output= -4*Return on Assets+20*Sales/ Total Assets-72.9*Cash from
Operations/Sales-10*EBIT/Interest
```

Although the evolved models were free to select from 22 potential explanatory variables, it is notable that each model only employed a small subset of these.

Table 17.1. The classification accuracies reported for each of the three years prior to failure

Years Prior to Failure	In-Sample Accuracy	Out-Of-Sample Accuracy
1	85.9%	80%
2	82.8%	80%
3	75.8%	70%

This lends support to the proposition that many financial ratios have similar information content and that classification accuracy is not enhanced through the construction of models with a large number of these ratios. It is also notable that each model has (approximately) included one variable drawn from the four main categories of explanatory variables suggested in the corporate failure literature.

The risk factors suggested by each model differ somewhat and contain plausible findings. Examining the best classifier evolved for T1 suggests that risk factors include low return on assets, low retained earnings and a high ratio of total liabilities to total assets, which concords with financial intuition. Less obviously, a high ratio of inventory to net liquid assets (inventory+receivables+cash-payables) is also a risk factor, possibly resulting from depletion of cash or build-up of inventories as failure approaches. Risk factors for firms at T2 include low return on assets and a low ratio of earnings to interest costs. Less intuitive risk factors indicated are a low ratio of fixed assets to total assets and a high ratio of sales to total assets. The former could indicate firms with a lower safety cushion of saleable resources which could be sold to stave off collapse, the latter could be serving as a proxy variable for firms with rapid sales growth. Over-rapid sales growth can be a danger signal, indicating that management resources are being spread too thinly. Finally, risk factors indicated for firms at T3 include low return on assets, a low ratio of profit to interest charge, a low level of cash generated from operations and, as for T2, a high ratio of sales to total assets.

Although the models for each year are evolved separately, the general form of each model appears consistent with the hypothesis that there is a financial trajectory towards failure. Low profits and high interest payments as a percentage of profits in periods T3 and T2 indicate a firm in financial difficulties, with an erosion of the safety cushion provided by high levels of (saleable) fixed assets indicated in the risk factors at T2. The final year prior to failure sees additional risk factors indicated by high levels of debt and reducing cash balances/inventory build-up.

17.4 Discussion

GE successfully evolved useful rules for prediction of corporate failure with a performance equivalent to that reported in prior studies. In assessing the performance of the developed models, a number of caveats must be borne in mind. The premise underlying this case study (and all empirical work on corporate failure prediction) is that corporate failure is a process, commencing with poor management decisions, and that the trajectory of this process can be tracked using accounting ratios. This approach does have inherent limitations. It will not forecast corporate failure which results from a sudden environmental event. It is also likely that the explanatory variables utilised contain noise. Commentators [11, 200] have noted that managers may attempt to utilise creative accounting practices to manage earnings and/or disguise signs of distress.

This case could be extended in a number of directions. Non-financial qualitative explanatory variables or variables related to the firm's share price could be incorporated to further improve classification accuracy. One advantage of this approach is that the model would incorporate more up-to-date data, as information drawn from financial statements is, by definition, lagged information.

Another interesting extension of the case would be to allow GE to evolve a good set of ratios for potential inclusion in the classification model, from the raw underlying financial information. To date, most attempts at developing models for the prediction of corporate failure have utilised a limited set of financial ratios. These ratios are generally selected on an ad hoc basis by the modeller due to the lack of a strong theoretical framework underlying the failure prediction problem. Unfortunately, the number of ratios which can be calculated from a set of financial statements is large. A set of financial statements could contain several hundred numbers between the primary financial statements and the detailed notes accompanying the primary statements, resulting in many thousands of potential financial ratios. Most studies in the corporate failure domain utilise similar financial ratios, circularly justifying the choice of ratios by reference to earlier studies. This methodological approach leaves open the possibility that alternative, better, representations of the financial data (ratios) exist.

To provide an example of the potential of this methodology, consider the following three grammars, where var 1 to var 12 are pieces of raw financial information such as sales or net profit before tax and interest:

```
Grammar 1

<lc> ::= output = <coeff> * ( ( <var> ) / ( <var> ) );

<var> ::= var1[index]  | var2[index]  | var3[index]
        | var4[index]  | var5[index]  | var6[index]
        | var7[index]  | var8[index]  | var9[index]
        | var10[index] | var11[index] | var12[index]

<coeff> ::=  ( <coeff> ) <op> ( <coeff> ) | <float>

<op> ::= + | -

<float> ::= 20 | -20 | 10 | -10 | 5 | -5
          | 4 | -4 | 3 | -3 | 2 | -2 | 1
          | -1 | .1 | -.1
```

```
Grammar 2

<lc> ::= output = <expr> ;

<expr> ::= ( <expr> ) + ( <expr> )
         | <coeff> * (<var>/<var>)

<var> ::= var1[index] | .... | var12[index]

<coeff> ::= (<coeff>) <op> ( <coeff> )
          | <float>

<op> ::= + | -

<float> ::= 20 | -20 | 10 | -10 | 5 | -5 | 4 | -4
          | 3 | -3 | 2 | -2 | 1 | -1| .1 | -.1


Grammar 3

<lc> ::= output = expr ;

<expr> ::= ( <expr> ) + ( <expr> )
         | <coeff> * ( <ratio> / <var> )

<ratio> ::= <ratio> <op> <ratio> | <var>

<var> ::= var1[index] | .... | var12[index]

<coeff> ::= ( <coeff> ) <op> ( <coeff> ) | <float>

<op> ::= + | -

<float> ::= 20 | -20 | 10 | -10 | 5 | -5 | 4 | -4
          | 3 | -3 | 2 | -2 | 1 | -1 | .1 | -.1
```

Grammar 1 permits the construction of a predictive rule consisting of a single ratio, formed from any two discrete pieces of raw financial data. This ratio can be rescaled as required by an evolved coefficient parameter. In essence, this grammar searches for the best univariate predictive model. Grammar 2 permits the construction of predictive rules which chain ratios together, producing linear rules of the form:

```
output = coefficient * Ratio X + coefficient * Ratio Y + ...
```

In each of these grammars, only ratios of the form $\frac{a}{b}$, where a and b are discrete pieces of financial data, are permitted.

Grammar 3 allows the construction of a linear chain of ratios, where the ratios can take the form $\frac{a+b+\cdots}{x}$, greatly increasing the number of possible ratios that can be formed from the raw data. Interested readers are referred to [29, 31] for further details.

18

Corporate Failure Prediction Using an Ant Model

Studies of the workings of ant colonies have inspired a series of ant colony algorithms which can be used for optimisation (ant colony optimisation, *ACO*) or clustering tasks. Broadly speaking, in ant systems search or learning activities are distributed over ant-like agents, which, in a metaphorical and highly stylised way, mimic the behaviour of real ants. The communication between the ant agents in these models during the search process is not direct, instead they communicate indirectly by modifying the environment faced by each other. There is no single 'Ant Model', rather there exists a family of models, each inspired by a different aspect of ant behaviour. These models include those inspired by:

- ant-foraging (co-operative food-retrieval) behaviour,
- brood-sorting behaviour,
- cemetery formation behaviour, and
- co-operative transport.

As yet, there are few applications of these models to the domain of finance. This case study examines the usefulness of an ant-clustering system for the prediction of corporate failure, using the same dataset as the last chapter.

18.1 Background

Several clustering and classification algorithms have been inspired by the observation of various sorting activities in ant colonies. At their simplest, the sorting activities entail the picking up and depositing of items by ants, into clusters of like-items. Examples of this include the *brood-sorting* behaviour of the ant species *Leptothorax unifasciatus*, where ant larvae are sorted by size and clustered at the centre of the brood area in the colony [62], and *cemetery building*, where dead ants are collected from the colony and deposited together.

The *pick-drop* behaviour of real ants can be applied to inspire a clustering algorithm as follows. Assume a set of n-dimensional items which are to be clustered into 'like' groups, and that it is wished to produce a visualisation of the clustering process. Once a distance metric is defined, the remaining task is to uncover a projection from n-dimensional space onto a 2-d plane such that neighbouring vectors in n dimensions are also neighbours on the 2-d plane. Unfortunately, the nature of the projection which will best achieve this for a given dataset is usually unknown, and the objective of the ant (or other clustering) model is to uncover this projection.

To achieve this, the initial projection of data vectors onto the plane can be random, but the projection is then modified by a population of artificial ants which randomly traverse the 2-d plane. These ants are programmed to follow a simple pick or drop rule whereby they pick up objects (each of which corresponds to one of the data vectors) when there are no similar objects nearby, and drop an object they are carrying when there are similar objects nearby. Through the pick-drop process, the ants act to modify the initial projection into one which, as far as possible, is topology preserving. The overall effect of the process is to sort similar objects (n-dimensional vectors) into clusters on the 2-d plane.

18.2 Methodology

As in the last case, a total of 178 firms were selected. For every failing firm, a matched non-failing firm is selected. Failed and non-failed firms are matched both by industry sector and size (sales revenue three years prior to failure). The set of 89 matched firms are randomly divided into model building (74 pairs) and out-of-sample (15 pairs) datasets. The dependent variable is binary (0,1), representing either a non-failed or a failed firm. For this case, the financial ratios chosen were guided by those selected by Altman [3], with the substitution of a ratio which required a market value measure, with cash from operations/total liabilities. The five ratios utilised were:

   i. R1 = Working Capital/Total Assets
  ii. R2 = Retained Earnings/Total Assets
 iii. R3 = Earnings before Interest and Tax/Total Assets
 iv. R4 = Cash from operations/Total Liabilities
  v. R5 = Sales/Total Assets

R1: This ratio is a measure of the net liquid assets of the firm relative to firm size.
R2: This ratio serves as a proxy for the trading history/age of the firm. A young firm will be likely to show a low RE/TA ratio, as it will not have had time to build up its retained earnings.

R3: This ratio measures the productivity of the firm's assets, removing any tax or leverage factors. A firm's existence, barring catastrophic environmental events, ultimately depends on its ability to earn a return on its assets.

R4: This ratio indicates the level of cash generation by the firm in a financial year, relative to the size of its liabilities.

R5: This ratio represents the efficiency with which the firm can generate sales, given its asset base. The ratio also serves as a proxy for the industrial sector of the firm, as well as the scale of production capacity which the firm has chosen to adopt.

18.2.1 Ant System

The clustering model used in this case is based on those of Deneubourg et al. [62] and Lumer and Faieta [146]. The idea behind these models is that objects should be picked up if they are not already beside similar objects. They should then be relocated and dropped beside other items of the same type.

Deneubourg Model

Under the original cemetery formation model of Deneubourg et al. [62], ants traverse an $x * y$ 2-d grid, randomly moving from their current site to one of four neighbouring sites (up-down-left-right) at each iteration of the algorithm. If an unladen ant encounters a corpse, it picks it up with probability P_{pick}, and in subsequent iterations may drop the corpse with probability P_{drop}. Assuming there is only one type of object in the environment the pick and drop probabilities can be defined as:

$$P_{pick} = \left(\frac{k_1}{k_1 + f}\right)^2 \tag{18.1}$$

$$P_{drop} = \left(\frac{f}{k_2 + f}\right)^2 \tag{18.2}$$

where f is the perceived fraction of all the objects in the neighbourhood of the ant (providing an estimate of the local density of dead ants or equivalently, an estimate of the size of the local cluster), and k_1 is a threshold constant.

When an ant encounters a corpse, and $f \ll k_1$, the ant is not considered to be in the vicinity of a large cluster, and therefore should pick up the corpse in order to drop it on a cluster somewhere else (therefore, P_{pick} should be close to one). Conversely, if an ant encounters a corpse and $f \gg k_1$, the ant is close to a large existing cluster and should not move the corpse as it is already in a large cluster of corpses (therefore P_{pick} should be close to zero).

The probability that a randomly moving loaded ant, drops an object (P_{drop}) is governed by a second threshold constant k_2. When $f \ll k_2$, the ant carrying a corpse is not close to a cluster of other corpses, and therefore should continue to carry the corpse until a cluster is found (P_{drop} is close to zero).

The value of f is an important parameter in the algorithm, as it directly impacts on both the probability of picking up or depositing a corpse. In the Deneubourg et al. algorithm, the value of f is calculated by each ant, based on its personal history. It is assumed that each ant has a T period memory. If we assume that an ant can only encounter 0 or 1 objects per time unit, and letting N represent the total number of objects encountered during T time periods, f is calculated as $= N/T$.

The algorithm leads to ant behaviour such that small clusters of dead ants (perhaps of size 1) are emptied, and large clusters grow. In turn, the large clusters will tend to merge. Extending this algorithm to cases where there is more than one object type, f is replaced by a series of f values, each representing the fraction of each type of object encountered during the last T time units [23].

Lumer and Faieta Model

The canonical model above was generalised by Lumer and Faieta [146], with the inclusion of a distance or *dissimilarity* measure between objects. Let $d(o_i, o_j)$ be the distance between two objects o_i and o_j in the space of object attributes. Assume that an ant is located at site r on a 2-d grid at time t, and it finds an object o_i at that site. The *local density* ($f(o_i)$) with respect to object o_i at site r is given by:

$$f(o_i) = max \left\{ 0, \frac{1}{s^2} \sum_{o_j \in Neigh_{(s*s)}(r)} \left[1 - \frac{d(o_i, o_j)}{\alpha} \right] \right\} \tag{18.3}$$

$f(o_i)$ (an analogue to f in the model of Deneubourg et al.) is a measure of the average similarity of o_i with other objects which are present in its neighbourhood, defined as the $s * s$ positions on the 2-d grid around the grid location of o_i which the ant can 'see'.[1] α is a *tuning knob* for the degree of dissimilarity discrimination between objects. If α is large, even quite dissimilar items may be clustered together, if it is small, distances between vectors in the attribute space are amplified, and even similar vectors may end up in different clusters.

Taking two extreme cases to demonstrate the calculation of local density, if all the sites around o_i are occupied by objects which are similar to it then $f(o_i)=1$ and o_i should be picked up with a low probability. If all sites around o_i are occupied by objects which are very dissimilar to it then $f(o_i)$ is small and o_i should be picked up with a high probability. Under the Lumer and

---

[1] Therefore, in comparison with the Deneubourg algorithm which uses a memory to calculate f, the LF algorithm allows each ant have a direct perception of the area surrounding its current location.

Faieta model, the pick and drop probabilities of the Deneubourg et al. model are altered to:

$$P_{pick}(o_i) = \left(\frac{k_1}{k_1 + f(o_i)} \right)^2 \tag{18.4}$$

$$P_{drop}(o_i) = 2f(o_i), \text{ if } f(o_i) < k_2 \tag{18.5}$$

$$P_{drop}(o_i) = 1, \text{ if } f(o_i) \geq k_2 \tag{18.6}$$

Algorithm Used in This Case Study

The algorithm used in this case study is closely modelled on that of Lumer and Faieta. Each company (i) is defined by a vector of its financial ratios, $Company_i = (r_1, ..., r_n)$, where n is the number of financial ratios used in the clustering model. In this case we limit n to 5.

Each company (i) is symbolised by an object o_i on the 2-d grid. Initially, these objects are randomly scattered over the 2-d grid, and during the execution of the algorithm they are clustered into heaps of similar items. The distance between two objects is calculated by the Euclidean distance between the two companies' vector of financial ratios in R^n. There is no direct link between the position of an object on the 2-d plane and its vector in R^n.

At the start of the algorithm, a fixed number of ants are placed on the 2-d grid. During each iteration of the algorithm, an ant may either be carrying an object or not. In the first case, the ant may:

- drop the object on a neighbouring empty location,
- drop the object on a neighbouring object, if both are similar, or
- drop the object on a neighbouring heap, if the object is similar to other members of the heap.[2]

If the ant is not already carrying an object, it may:

- pick up a single object from a neighbouring location, or
- pick up the most dissimilar object from a heap on a neighbouring location.

The pseudo-code for the algorithm follows:

```
{Initialization} For every company Oi do
       Place Oi randomly on grid
End For

For all ants do
       Place ant at randomly selected site
End For

{Main Loop} For t = 1 to tmax  do
       For all ants do
       IF((ant unloaded)and(site occupied by company Oi)) then
```

[2] A heap arises when there are multiple objects on a single grid location.

```
      Compute f(Oi) and Prob (pick)(Oi)
      Draw random real number R between 0 and 1
      IF (R <=Prob (pick) (Oi) then
          Pick up company Oi
      End If
      ELSE If((ant carrying company Oi)
          and (site is empty)) then
      Compute f(Oi) and Prob (drop)(Oi)
      Draw random real number R between 0 and 1
      IF ( R<= Prob (drop)(Oi) ) then
          Drop company Oi
      End If
      End If
   Move to randomly selected neighbouring site not
      occupied by other ant
   End For
End For
```

The algorithm acts to construct clusters, such that the distances between objects of the same cluster are small in comparison with the distances between objects in different clusters. As the algorithm runs, and clusters start to form, the probability of objects being picked up diminishes and $\lim_{t \to \inf} Prob_{pick} \to 0$, as similar objects are grouped together.

Fig. 18.1. As the algorithm iterates the individual points representing failed and healthy companies group into distinct clusters

Analysing the Output from the Algorithm

The output from the clustering algorithm is a visual map of the individual firms (Fig. 18.1). Like self-organising maps (SOMs) the ant clustering algorithm is unsupervised in that it does not make use of a priori group memberships during the training process. Therefore, once the training process is complete and a number of clusters have been created by the algorithm, the modeller must assign a class label to each cluster. The simplest way to do this is to assign a label to each cluster based on whether the companies in that cluster are primarily healthy or failed. Once the label is assigned, all companies in the cluster are given the same classification. By comparing these classifications with the known classifications for the training companies, the in-sample accuracies for the model can be obtained.

The out-of-sample data can then be classified by determining which cluster each out-of-sample item is closest to. A basic method of doing this is to calculate the mean values on each dimension for all the training items in a cluster. Once the mean value for the 'centre' of each cluster has been determined, every out-of-sample data vector is assigned to the cluster to which it is closest using Euclidean distance, and the out-of-sample classification accuracy is then determined.

18.3 Results

The efficiency of the algorithm and the number of clusters which are identified in the data depends on the choices of the parameters for the algorithm which the user selects. For example, choosing a large grid size will tend to increase the run-time of the algorithm as the ants spend much time traversing empty grid positions. A large grid size will also tend to produce a greater number of clusters, particularly in the earlier stages of the algorithm. Using too few ants also results in a very slow clustering process. Following a trial and error process to determine the parameter settings for the ant system, the final configuration was as follows:

- a population of 40 ants,
- k_1=0.12,
- k_2=0.3,
- α=1.15,
- $step$=6 (distance travelled on the grid by an ant during each iteration of the algorithm),
- t_{max}=3.5E+5 (number of iterations of algorithm), s=3, and
- grid size = 150*150.

These settings are broadly similar to those used by Lumer and Faieta. Accuracy of the developed models was assessed based on their classification accuracy on the out-of-sample dataset. Summarised classification accuracies for the

Fig. 18.2. A screenshot from the ant clustering system showing a number of clusters, and highlighting one of the out-of-sample datapoints

ant models for the three years prior to failure and for a benchmark LDA model are provided in Table 18.1. Additional metrics were collected on the positive accuracy (correct prediction of non-failure) and negative accuracy (correct prediction of failure) for each of the models. Table 18.2 provides these for the out-of-sample datasets for the LDA and ant models.

Table 18.1. LDA vs ant model out-of-sample results for one to three years prior to failure

Results	LDA	Ant
T-1	78.0%	66.67%
T-2	58.0%	73.33%
T-3	58.0%	56.67%

The overall results indicate that both the ant and the LDA models can correctly distinguish between solvent and insolvent firms. The classification accuracy of the LDA model exceeds that of the ant model in both T-1 and T-3, with the ant model outperforming the LDA model in T-2.

Table 18.2. Out-of-sample positive and negative classification accuracy for LDA and ant models, one to three years prior to failure

LDA	Positive Accuracy	Negative Accuracy
T-1	80.0%	76.0%
T-2	76.0%	40.0%
T-3	52.0%	64.0%

Ant		
T-1	80.0%	53.33%
T-2	80.0%	66.67%
T-3	46.67%	66.67%

Table 18.3. Stability of results of ant model for T-1

Recut	Positive Accuracy	Negative Accuracy	Overall
1	86.66%	53.33%	70.00%
2	80.00%	53.33%	66.66%
3	60.00%	66.66%	63.33%
4	73.33%	53.33%	63.33%
5	60.00%	40.00%	50.00%
6	80.00%	40.00%	60.00%
7	86.66%	33.33%	60.00%
8	86.66%	20.00%	53.33%
9	80.00%	60.00%	70.00%
10	73.33%	66.66%	70.00%

Prior studies of corporate failure have suggested that it is more difficult to accurately predict insolvency than solvency, as firms can fail for a multitude of reasons, not all of which result from a failure trajectory [158]. Therefore, the overall results are decomposed to examine the positive and negative classification accuracies (correct prediction of non-distressed, and correct prediction of distressed) for each model. The expected pattern emerges for the LDA model, and for the ant model in T-1 and T-2.

To assess the stability of the results from the ant model for different randomisations of the training/out-of-sample data, ten recuts of the dataset were performed, and the out-of-sample classification accuracies for each of these are presented in Table 18.3. The classification accuracies display variability, but we are unable to categorically state whether this arises due to the stochastic nature of the algorithm or because of the small size of the out-of-sample dataset. Further analysis with a larger dataset is required to cast more light on the stability of the algorithm's performance.

18.4 Discussion

The objective of this case is to demonstrate a novel application of an ant-inspired clustering model, the prediction of the solvency of corporations. The results suggest that corporate failure can be forecast using an ant methodology, and also suggest that ant-inspired algorithms could be useful for predicting credit ratings.

Several opportunities exist to extend this case. We did not attempt to optimise the selection of ratio inputs, or to consider a wider range of financial/non-financial inputs on each firm or the environment, and this step could have improved the results of the ant model and the benchmark LDA model. It is also possible that the results for the ant model could have been improved if an exhaustive search to find the optimal parameters of the ant model had been undertaken. There are several other ant-clustering algorithms in existence [180, 183], and it would be interesting to compare the performance of each of these algorithms on this and other financial classification problems.

19

Bond Rating Using Grammatical Evolution

Most large firms use both share and debt capital to provide long-term finance for their operations. The debt capital may be raised from a bank loan, or may be obtained by selling bonds directly to investors. As an example of the scale of US bond markets, the value of new bonds issued in 2004 totaled $5.48 trillion, and the total value of outstanding marketable bond debt at 31 December 2004 was $23.6 trillion [24].[1]

When a company wants to issue traded debt (bonds), it must obtain a credit rating for the issue from at least one recognised rating agency (Standard & Poor's (S&P), Moody's, Fitches' or Dominion Bond Rating Service). The credit rating represents the rating agency's opinion at a specific date of the creditworthiness of a borrower in general (an *issuer credit rating*), or in respect of a specific debt issue (a *bond credit rating*). These ratings impact on the borrowing cost, and the marketability of issued bonds.

In common with the related corporate failure prediction problem, a feature of the bond-rating prediction problem is that there is no clear theoretical framework for guiding the choice of explanatory variables, or model form. In the absence of an underlying theory, most published work on credit rating prediction employs a data-inductive modelling approach, using firm-specific financial data as explanatory variables, in an attempt to recover the rating model used by an agency. This produces a high-dimensional combinatorial problem, as the modeller is attempting to uncover a good set of explanatory variables, and model form, giving rise to particular potential for an evolutionary automatic programming methodology such as grammatical evolution (GE). This case demonstrates the application of GE in order to construct a model which can predict the bond rating of a firm.

---

[1]In comparison, the total *global* market capitalisation of all companies quoted on the NYSE at 31/12/04 was $19.8 trillion [164].

19.1 Background

Several categories of individuals would be interested in a model that could produce accurate estimates of bond ratings. Such a model would be of interest to firms that are considering issuing debt as it would enable them to estimate the likely return investors would require if the debt was issued, thereby providing information for pricing the bonds. The model could also be used to assess the creditworthiness of firms that have not issued debt and hence do not already have a published bond rating. This information would be useful to bankers or other companies that are considering whether they should extend credit to that firm.

Most rated debt is publicly tradable on stock markets, and bond ratings are typically changed infrequently. An accurate bond-rating prediction model could indicate whether the current rating of a bond is still justified. To the extent that an individual investor could predict a bond re-rating before other investors foresee it, this may provide a trading edge.

In addition, the recent introduction of credit-risk derivatives allows investors to buy protection against the risk of the downgrade of a bond [5]. The pricing of such derivative products requires a quality model for estimating the likelihood of a credit rating change.

Notation for Credit Ratings

Although the precise notation used to denote the creditworthiness of a bond or issuer varies between rating agencies, the credit status is generally denoted by means of a discrete, mutually exclusive, letter rating. Taking the rating structure of S&P as an example, the ratings are broken down into 10 broad classes. The highest rating is denoted AAA, and the ratings then decrease in the following order, AA, A, BBB, BB, B, CCC, CC, C, D. Ratings between AAA and BBB (inclusive) are deemed to represent *investment grade*, with lower quality ratings deemed to represent debt issues with significant speculative characteristics (also called *junk bonds*). A 'C' grade represents a case where a bankruptcy petition has been filed, and a 'D' rating represents a case where the borrower is currently in default on their financial obligations. As would be expected, the probability of default depends strongly on the initial rating which a bond receives (Table 19.1). Ratings from AAA to CCC can be modified by the addition of a + or a - to indicate at which end of the rating category the bond rating falls.

19.1.1 Rating Process

Rating agencies earn fees from bond issuers for evaluating the credit status of new issuers and bonds, and for maintaining credit rating coverage of these firms and bonds. A company obtains a credit rating for a debt issue by contacting a rating agency and requesting that an issue rating be assigned to the

Table 19.1. Rate of default by initial rating category (1987-2002) (from Standard & Poor's, 2002)

Initial Rating	Default Rate (%)
AAA	0.52
AA	1.31
A	2.32
BBB	6.64
BB	19.52
B	35.76
CCC	54.38

new debt to be issued, or that an issuer rating be assigned to the company as a whole. As part of the process of obtaining a rating, the firm submits documentation to the rating agency including recent financial statements, a prospectus for the debt issue, and other non-financial information. Discussions take place between the rating agency and management of the firm and a rating report is then prepared by the analysts examining the firm. This rating report is considered by a rating committee in the rating agency which decides the credit rating to be assigned to the debt issue/issuer.

Rating agencies emphasise that the credit rating process involves consideration of financial as well as non-financial information about the firm, and also considers industry and market-level factors. The precise factors and related weighting of these factors used in determining a bond's rating are not publicly disclosed by the rating agencies. Subsequent to their initial rating, a bond may be re-rated upwards (upgrade) or downwards (downgrade) if company or environmental circumstances change. A re-rating of a bond below investment grade to junk bond status (such bonds are colourfully termed *fallen angels*) may trigger a significant sell-off as many institutional investors are only allowed, by external or self-imposed regulation, to hold bonds of investment grade.

19.2 Methodology

The dataset consists of financial data drawn from the financial statements of 791 non-financial US companies, along with their associated S&P bond-issuer credit-rating. In this case, we restrict attention to discriminating between investment grade vs junk grade ratings. In the dataset 57% of companies have an investment-grade rating (AAA, AA, A or BBB), and 43% have a junk rating. To allow time for the preparation of year-end financial statements, the filing of these statements with the Securities and Exchange Commission (SEC), and the development of a bond rating opinion by Standard & Poor's rating agency, the bond rating of the company as at 30 April 2000, is matched with financial information drawn from their financial statements as at 31 December 1999.

Sample Selection

A subset of 600 firms was randomly sampled from the total of 791 firms to produce two groups of 300 investment grade and 300 junk rated firms. The 600 firms were randomly allocated to the training set (420) or the hold-out sample (180), ensuring that each set was equally balanced between investment and non-investment grade ratings.

Input Selection

A total of eight financial variables were selected for inclusion in this study. The selection of these variables was guided both by prior literature in bankruptcy prediction, literature on bond rating prediction, and by preliminary statistical analysis. The financial ratios chosen during the selection process were:

   i. Current ratio
  ii. Retained earnings to total assets
 iii. Interest coverage
  iv. Debt ratio
   v. Net margin
  vi. Market to book value
 vii. Total assets
viii. Return on total assets

The objective in selecting a set of proto-explanatory variables is to choose financial variables that vary between companies in different bond rating classes, and where information overlaps between the variables are minimised. Comparing the means of the above ratios for the two groups of ratings (Table 19.2), reveals a statistically significant difference at the 1% level, and, as expected, the financial ratios in each case for the investment ratings are stronger than those for the junk ratings. The only exception is the current ratio, which is stronger for the junk rated companies, possibly indicating a preference for these companies to hoard short-term liquidity, as their access to long-term capital markets is weak. Table 19.3 provides a correlation analysis between the selected ratios. Examination of the correlations indicates that most are quite low, with the only notable correlation being between the debt ratio and (retained earnings/total assets) ratio.

Grammar

The grammar adopted in this study is as follows:

Table 19.2. Means of input ratios for investment and junk bond groups of companies

	Investment grade	Junk grade
Current ratio	1.354	1.93
Retained earnings/Total assets	0.22	-0.12
Interest coverage	7.08	1.21
Debt ratio	0.32	0.53
Net margin	0.07	-0.44
Market to book value	18.52	4.02
Total assets	10083	1876
Return on total assets	0.10	0.04

Table 19.3. Correlations between financial ratios

	CR	RE/TA	IC	DR	NM	MTB	TA	ROA
CR	1	-0.08	-0.01	0.06	-0.27	0.01	-0.18	-0.15
RE/TA	-0.08	1	0.27	-0.64	0.14	0.15	0.15	0.48
IC	-0.01	0.27	1	-0.28	0.06	0.31	0.15	0.41
DR	0.06	-0.64	-0.28	1	-0.05	-0.19	-0.20	-0.27
NM	-0.27	0.14	0.06	-0.05	1	0.01	0.03	0.22
MTB	0.01	0.15	0.31	-0.19	0.01	1	0.04	0.14
TA	-0.18	0.15	0.15	-0.20	0.03	0.04	1	0.07
ROA	-0.15	0.48	0.41	-0.27	0.22	0.14	0.07	1

```
<lc> ::= if( <expr> <relop> <expr> ) class=''Junk''; else class=''Investment Grade'';

<expr> ::= ( <expr> ) + ( <expr> ) | <coeff> * <var>

<var> ::= var3 | var4 | var5 | var6 | var7 | var8 | var9 |var10 | var11

<coeff> ::= ( <coeff> ) <op> ( <coeff> ) | <float>

<op> ::= + | - | *

<float> ::= 9 | 8 | 7 | 6 | 5 | 4 | 3 | 2 | 1 | -1 | .1

<relop> ::= <=
```

where **var3** = Current Ratio, **var4** = Retained Earnings to total asset, **var5** = Interest Coverage, **var6** = Debt Ratio, **var7** = Net Margin, **var8** = Market to book value, **var9** = Total Assets, **var10** = ln (Total Assets), **var11** = Return on total assets.

19.3 Results

Each of the GE experiments is run for 100 generations, with variable-length one-point crossover at a probability of 0.9, one point bit mutation at a probability of 0.01, roulette selection, and steady-state replacement. Results are reported for two population sizes (500 and 1000), and for two distinct fitness functions. To assess the stability of the results across different randomisations of the dataset between training and test data, the dataset was recut five times, maintaining an equal balance of investment and non-investment grade ratings in the resulting training and test datasets.

In the experiments, fitness is defined as the number of correct classifications produced by an evolved discriminant rule. The results for the best individual of each cut of the dataset, where 30 independent runs were performed for each cut, averaged over all five randomisations of the dataset, for both the 500 and 1000 population sizes, are given in Table 19.4.

Table 19.4. Performance of the best evolved rules on their training and out-of-sample datasets, averaged over all five randomisations

	Fitness	TP	TN	FP	FN
Train GEBOND500	86.15%	185.2	175.8	33.4	24.6
Train GEBOND1000	86.78%	183.2	180.4	28.8	26.6
Out-Sample GEBOND500	85.60%	77.6	75.8	13.6	12.2
Out-Sample GEBOND1000	86.26%	77.8	76.6	12.6	12

To assess the overall hit-ratio (classification accuracy) of the developed models (out-of-sample), Press's Q statistic was calculated for each model. In all cases, the null hypothesis, that the out-of sample classification accuracies are not significantly better than those that could occur by chance alone, was rejected at the 1% level. A t-test of the hit-ratios also rejected a null hypothesis that the classification accuracies were no better than chance at the 1% level. Across all the data recuts, the best individual achieved an 87.56 (84.36)% accuracy in-sample (out-of-sample) when the population size was 500, with the best individual across all data recuts in the population=1000 case obtaining an accuracy of 87.59 (84.92)% accuracy in-sample (out-of-sample). Although the average out-of-sample accuracy obtained for population=1000 slightly exceeds that for population=500, the difference was not found to be statistically significant. A plot of the best and average fitness on each cut of the in-sample dataset, for the population=500 case, can be seen in Fig. 19.1, and for the case where population=1000 in Fig. 19.2.

A second series of experiments was undertaken using a fitness measure that takes into consideration negative classifications. The fitness measure adopted earlier does not explicitly take into consideration the amount of under and over-prediction represented by false negatives and false positives, respectively. A more general measure considers the correlation between the prediction and the observed reality [148]. A correlation measure indicates how much better a particular predictor is than random predictions, and has been adopted previously in GP [132]. In the case of a binary classification problem the correlation measure, C, and the calculation of a corresponding fitness value, are given below.

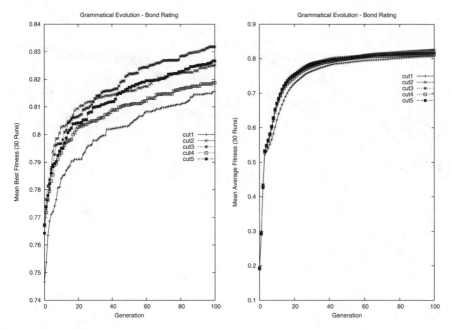

Fig. 19.1. Best and average fitness values of 30 runs on the five recuts of the in-sample dataset with a population size of 500

$$C = \frac{(N_{tp} * N_{tn} - N_{fn} * N_{fp})}{\sqrt{(N_{tn} + N_{fn}) * (N_{tn} + N_{fp}) * (N_{tp} + N_{fn}) * (N_{tp} + N_{fp})}} \quad (19.1)$$

$$Fitness = \frac{(C+1)}{2} \quad (19.2)$$

where N_{tp}, N_{tn}, N_{fp}, N_{fn} are the number of true positives, true negatives, false positives, and false negatives respectively. A fitness value of zero means there is no correlation to the observed cases, a value of 0.5 means the classification accuracy is no better than random, and a value of 1.0 means a perfect correlation to observed cases. Results using this fitness measure for population sizes of both 500 and 1000 are provided in Table 19.5. Assessing the out-of-sample hit-ratio of the developed models using Press's Q statistic rejects the null hypothesis, that the out-of sample classification accuracies are not significantly better than those that could occur by chance alone, at the 1% level, however the models developed using this fitness function did not outperform those developed using the initial fitness function. As for the initial fitness function, the average classification accuracy (out-of-sample) was slightly higher for the case where population=1000 than for population=500, but the difference was not statistically significant.

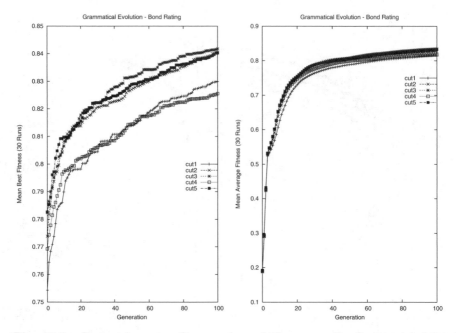

Fig. 19.2. Best and average fitness values of 30 runs on the five recuts of the in-sample dataset with a population size of 1000

Table 19.5. Performance of the best evolved rules on their training and out-of-sample datasets, averaged over all five randomisations

	Fitness	TP	TN	FP	FN
Train GEBOND500	0.8568	183.8	175.2	34	26
Train GEBOND1000	0.8644	187.8	174.4	34.8	22
Out-sample GEBOND500	0.8033	66	77.8	11.4	23.8
Out-sample GEBOND1000	0.8514	78.8	73.6	15.6	11

A plot of the best and average fitness on the five recuts of the in-sample dataset can be seen in Fig. 19.3 for population=500, and Fig. 19.4 for population=1000.

Structure of the Evolved Classification Rules

Examining the structure of the best individual in the case where the initial fitness function was utilised and where population=500 shows that the evolved discriminant function had the following form:

IF $(10+16\text{var}6-9\text{var}4-2\text{var}9) \geq 0$ THEN 'Junk' ELSE 'Investment Grade'

where var6 is *Debt Ratio*, var4 is $\frac{Retained\ Earnings}{Total\ Assets}$, and var9 is *Total Assets*.

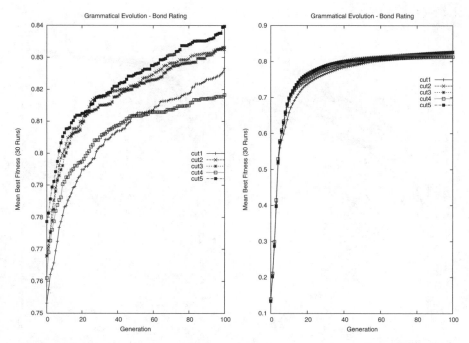

Fig. 19.3. Best and average fitness values of 30 runs on the five recuts of the in-sample dataset using the correlation fitness measure with a population size of 500

In the case where population=1000, the best evolved discriminant function had a similar form to the above:

IF (5+8var6-4var4-var9) ≥ 0 THEN 'Junk' ELSE 'Investment Grade'

Examining the signs of the coefficients of the evolved rules does not suggest that they conflict with common financial intuition. The rules indicate that low/negative retained earnings, low/negative total assets or high levels of debt finance are symptomatic of a firm that has a junk rating. Conversely, low levels of debt, a history of successful profitable trading, and high levels of total assets are symptomatic of firms that have an investment grade rating.

19.4 Discussion

The objective of this case was to illustrate the application of GE to model the corporate bond rating process. Despite using data drawn from companies in a variety of industrial sectors, the developed models showed an impressive capability to discriminate between investment and junk rating classifications. Several extensions of the case study are possible. The predictive target could

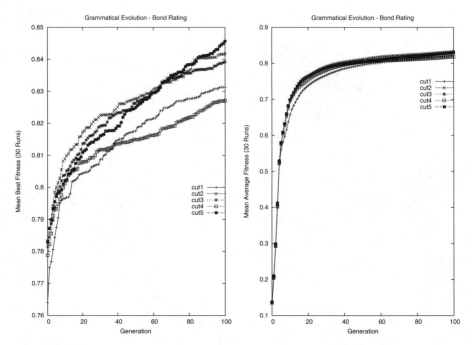

Fig. 19.4. Best and average fitness values of 30 runs on the five recuts of the in-sample dataset using the correlation fitness measure with a population size of 1000

be changed to predict the precise bond rating (a multi-classification model), or to predict which bonds will be re-rated in the foreseeable future. Another extension would be to extend the range of data inputs used in the model to include non-financial information about the firm and its industrial sector, or to include measures of how the firm is performing relative to its industry peers. Another possibility, particularly if the object is to anticipate a rerating, is to include metrics on changes in the financial ratios of the firm over the past few years, for example, have profits declined over the past three years, or have debt levels increased?

Bond Rating Using AIS

The natural immune system is a highly complex system, comprised of an intricate network of specialised tissues, organs, cells and chemical molecules. The natural immune system can recognise, destroy and remember an almost unlimited numbers of pathogens (foreign objects that enter the body, including viruses, bacteria, multi-cellular parasites, and fungi). To assist in protecting the organism, the immune system has the capability to distinguish between self and non-self. Notably, the system does not require exhaustive training with negative (non-self) examples to make these distinctions, but can identify items as non-self which it has never before encountered. The mechanisms of natural immune systems, including their ability to distinguish between self and non-self proteins, provide a rich metaphorical inspiration for the design of pattern recognition algorithms. This chapter illustrates how an algorithm drawn from an immune system metaphor can be used to create a classification system to distinguish between investment and junk rated bonds.

20.1 Methodology

The same dataset is used in this case as in the last chapter, and consists of financial information on 791 US companies, drawn from the S&P Compustat database. Again, the S&P bond rating of each company as at 30 April 2000 is matched with financial information drawn from their financial statements as at 31 December 1999. As in the last chapter, the eight financial ratios used are:

   i. Current ratio
  ii. Retained earnings to total assets
 iii. Interest coverage
 iv. Debt ratio
  v. Net margin
 vi. Market to book value
vii. Total assets

viii. Return on total assets

All ratios in the dataset were normalised into the range (0,1) after clipping outliers. Therefore the self /non-self space is an eight-dimensional hypercube, where each dimension corresponds to an individual financial ratio. In order to train the detectors, 25% of the self-data (companies with investment-grade ratings) was randomly chosen.

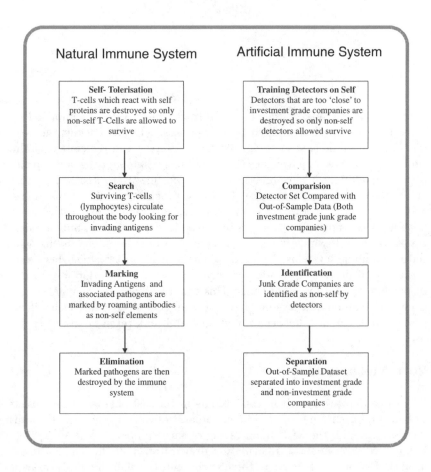

Fig. 20.1. Comparison of natural immune system processes and the negative selection algorithm for bond rating classification

The Negative Selection Algorithm

The canonical real-valued negative selection algorithm is compared to processes in the natural immune system in Fig. 20.1. Initially a predetermined number of detectors are randomly created. During the training, or tolerisation, process any detector that falls within a threshold distance r_s of any elements of the training set of self samples is discarded. The process of detector generation is iterated until the required number of valid detectors is created. The pseudo-code for the algorithm is as follows (S is the set of self-samples, r_s is a predefined threshold distance, and it is assumed that the search-space is bounded by an n-dimensional (0,1) hypercube):

   i. Detector set (D) is empty
  ii. Repeat
 iii. Create a random vector $\mathbf{x}$, drawn from $[0, 1]^n$
  iv. For for every $\mathbf{s_i}$ in $S, \mathbf{s_i} : i = 1, 2, ..., m$
   v. Calculate the Euclidean distance (d) between $\mathbf{s_i}$ and $\mathbf{x}$
  vi. If $d \leq r_s$ go to step ii
 vii. Add $\mathbf{x}$ (a valid non-self detector) to set D
viii. Until D contains the required number of valid detectors

Once the required number of detectors has been created, they can be used to classify new data observations. To do this the new data vector is presented to the population of detectors, and if it does not fall within r_s of any of them, the data vector is deemed to be self, as it did not trigger any of the non-self detectors, otherwise the new data vector is deemed to be non-self. In this case study, self is defined as a company with an investment grade rating, and each company is characterised by its vector of eight accounting ratios.

Algorithm Design

When developing the classification system, the non-self detectors are initially located randomly in the self/non-self space. The distance between each detector and each self company in the training sample being used to construct the classifier is calculated, and if a detector is within the threshold distance of any self vector, it is discarded and a new detector is randomly created to replace it. A detector is discarded where the median distance between it and the k-nearest self vectors is less than the threshold distance. The use of k-nearest neighbours (rather than just the nearest self vector) makes the algorithm less susceptible to noise in the input data [96]. In our experiments k is set at 5.

Once the full population of valid detectors has been generated, they can be exposed to new (out-of-sample) data, and used to predict whether these companies have an investment grade rating or not. New data vectors which are within range of a detector, and therefore which have characteristics similar to non-investment grade companies, are classed as having a non-investment grade rating, otherwise the vector is classed as an investment grade company.

As in the training process, determining whether a detector has been triggered depends the Euclidean distance between the detector and a test data vector. A vector (company) in the out-of-sample test set that is within a threshold distance of any detector is deemed to be a non-self, a non-investment grade company.

Algorithm Settings

The algorithm requires that the modeller specifies both the number of detectors and the size of the threshold distance. Following experimentation on the dataset, the number of detectors was set at 500, and a threshold distance of 0.80 was applied. Intuitively, the number of detectors and the size of the threshold distance determines the degree of coverage of the self/non-self space by the detectors. As the number of detectors and their threshold distance increases, the model will tend to misidentify investment-grade companies as being non-investment grade. Conversely, if there are very few detectors, and/or they have a small threshold distance, the model will tend to misidentify non-investment grade companies as being investment grade. Hence, the selection of the number of detectors and the threshold distance seeks to trade-off these two errors.

20.2 Results

The overall classification accuracy on the out-of-sample datasets was used to assess the performance of the developed classification model. Averaged over three recuts of the dataset between training and test data, a classification accuracy of 70.86% was obtained. This accuracy is poorer than that obtained by the GE classifier developed in the last chapter, however it is still sufficient to reject a null hypothesis, at the 5% level, that the out-of-sample classification accuracies are not significantly better than chance.

20.3 Discussion

In this case a novel methodology inspired by the workings of the natural immune system, the negative selection algorithm, was introduced and applied for the purposes of prediction of corporate bond-issuer ratings. The developed classifier was found to be able to distinguish between firms with investment grade and non-investment grade ratings with reasonable accuracy, using a small number of financial ratios drawn from the financial statements of those companies.

Artificial immune systems represent a relatively new class of algorithms and, as yet, few business applications of these algorithms have been developed. The negative selection algorithm used in this case has general utility for classification and it can be applied in a wide variety of settings. Examples of potential business applications include corporate failure prediction [27], fraud detection, and customer database segmentation.

A number of improvements could be implemented to further improve the efficiency of the algorithm. It is not generally efficient to attempt to cover all the non-self space with detectors, rather the aim is to cover regions of non-self space in which future observations of non-self are more likely to occur. Unlike the scenario faced by natural immune systems, in this case historic examples of non-self (junk-rated companies) already exist. These could be used to seed the process of creating valid detectors in order to speed it up. The task of generating a population of valid detectors grows rapidly as the size of self increases. Therefore seeding could be particularly useful when developing classification systems for high-dimensional business applications. Another possibility is to automate the selection of parameters such as the number of detectors. Recent work by Gonzalez and Cannady [97] demonstrates the potential of a hybrid AIS system which embeds an evolutionary algorithm for the purposes of automating parameter selection.

Another extension to the methodology adopted in this case would be to apply the variable-size detector algorithm, which was developed by Ji and Dasgupta [117]. In the canonical real-valued negative selection algorithm described above, the detectors have a fixed radius of detection. In the variable detector algorithm, the radius of each detector is permitted to differ. This allows areas of non-self which are far removed from any self vectors to be covered with a relatively small number of large radius detectors, and also allows for the insertion of smaller detectors to cover any gaps or holes in the non-self space between the large detectors.

21

Wrap-up

The objective of this book has been to provide readers with an up-to-date introduction to a broad range of biologically inspired algorithms, an introduction to trading system design, and an illustration of the practical application of several of the biologically inspired algorithms introduced in the book. We hope that this book will help spark new ideas in the minds of readers to encourage them to undertake their own work in financial modelling using biologically inspired methodologies.

The nature of financial modelling is that it will always be a difficult domain for prediction. As markets respond quickly to unanticipated events, prices can never be completely predictable, and investing will always carry the risk of loss. The limitations on our ability to test models should also be borne in mind. Experiments in financial modelling cannot be replicated under controlled conditions, and historical values of market prices and fundamental data only provide a single sample path of the market's behaviour through time. It is unrealistic to suppose that a model with a limited number of explanatory variables will produce high-quality predictions indefinitely, or in all market conditions. Models will date and require replacement. To minimise the risk of catastrophic failure, traders must combine carefully crafted models with carefully crafted money-management strategies, which diversify investments across markets and across multiple trading models.

No claim can be made that the recent advances in biologically inspired algorithms described in this book provide an easy route to get-rich-quick trading systems. The new tools do not remove the traditional modelling requirements for domain knowledge, the careful selection and preprocessing of inputs, the postprocessing of outputs, and the all-important plausibility test. Financial markets are a battleground, and information collection and analysis technologies represent the weaponry of traders. Novel biologically inspired algorithms provide us with powerful new tools which are capable of detecting subtle patterns between inputs and outputs. Carefully applied these have the potential to provide at least a temporary trading edge.

References

1. Aickelin, U. and Cayzer, S. (2002). The Danger Theory and Its Application to Artificial Immune Systems, in *Proceedings of the 1st International Conference on Artificial Immune Systems*, pp. 141-148, Canterbury, UK.
2. Allen, F. and Karjalainen, R. (1999). Using genetic algorithms to find technical trading rules, *Journal of Financial Economics*, 51:245-271.
3. Altman, E. (1968). Financial ratios, discriminant analysis and the prediction of corporate bankruptcy, *Journal of Finance*, 23:589-609.
4. Altman, E. (1993). *Corporate Financial Distress and Bankruptcy*, New York: Wiley.
5. Altman, E. (1998). The importance and subtlety of credit rating migration, *Journal of Banking and Finance*, 22:1231-1247.
6. Altman, E. (2000). Predicting Financial Distress of Companies: Revisiting the Z-score and Zeta Models, `http://www.stern.nyu.edu/~ealtman/Zscores.pdf` (accessed October 2001).
7. Altman, E., Haldeman, R. and Narayanan, P. (1977). ZETA analysis: A new model to identify bankruptcy risk of corporations, *Journal of Banking and Finance*,1:29-54.
8. Anderson, T., Bollerslev, T. and Das, A. (2001). Variance-ratio statistics and high-frequency data: testing for changes in intraday volatility patterns, *Journal of Finance*, LVI(1):305-327.
9. Angel, J. (1997). Tick size, share prices, and stock splits, *Journal of Finance*, LII(2):655-680.
10. Angeline, P. (1998). Using selection to improve particle swarm optimization, in *Proceedings of the IEEE International Conference on Evolutionary Computation*, Anchorage, May 1998, pp. 84-89, IEEE Press.
11. Argenti, J. (1976). *Corporate Collapse: The Causes and Symptoms*, London: McGraw-Hill.
12. Back, B., Laitinen, T., Sere, K. and van Wezel, M. (1996). Chosing Bankruptcy Predictors Using Discriminant Analysis, Logit Analysis and Genetic Algorithms, *Technical Report No. 40, Turku Centre for Computer Science*, Turku School of Economics and Business Administration.
13. Baestanes, D.E., Van Den Bergh, W.M. and Wood, D. (1994). *Neural Network Solutions for Trading in Financial Markets*, London: Pitman Publishing.

14. Banzhaf, W. (1994). Genotype-phenotype-mapping and neutral variation – A case study in genetic programming, in Lecture Notes in Computer Science 866, *Parallel Problem Solving from Nature III*, pp. 322-332, Springer.

15. Banzhaf, W., Nordin, P., Keller, R.E., and Francone, F.D. (1998). *Genetic Programming – An Introduction: On the Automatic Evolution of Computer Programs and Its Applications*, Morgan Kaufmann.

16. Bauer R. (1994). *Genetic Algorithms and Investment Strategies*, New York:Wiley.

17. Beaver, W. (1966). Financial ratios as predictors of failure, *Journal of Accounting Research - Supplement: Empirical Research in Accounting*, 71-102.

18. Bellanta, J. and Kadlec, J. (1985). Introduction to immunology, in *Immunology: Basic Processes*, ed. Bellanti, J., pp. 1-15, Philadelphia: W.B. Saunders.

19. BIS (2001). Central Bank Survey of Foreign Exchange and Derivatives Market Activity in April 2001, *Press Release, 31/2001E, Bank of International Settlements*, published October 2001.

20. BIS (2004). Triennial Central Bank Survey of Foreign Exchange and Derivatives Market Activity in April 2004, *Press Release, Bank of International Settlements*, http://www.bis.org/press/p040928.htm, published September 2004.

21. Blackwell, T.M. and Bentley, P.J. (2002). Dynamic search with charged swarms, In *Proceedings o f the Genetic and Evolutionary Computation Conference-GECCO 2002*, Spector et al. (Eds.), New York, USA, July 9-13, 2002, pp. 19-26, Morgan Kaufmann.

22. Blackwell, T.M. and Branke, J. (2004). Multi-swarm optimization in dynamic environments, in LNCS 3005 *EvoWorkshops 2004*, Coimbra, Portugal, pp. 489-500, Springer.

23. Bonabeau, E., Dorigo, M. and Theraulaz, G. (1999). *Swarm Intelligence: From Natural to Artificial Systems*, Oxford: Oxford University Press.

24. Bond Market Association - Research Quarterly (2005). (February 2005) http://www.bondmarkets.com/assets/files/Research_Quarterly_0205.pdf.

25. Brabazon, A. (2002a). Neural network design using an evolutionary algorithm, *Irish Accounting Review*, 9(1):1-18.

26. Brabazon, A. (2002b). Financial time series modelling using neural networks: An assessment of the utility of a stacking methodology, in *Proceedings of AICS 2002*, Lecture Notes in Artificial Intelligence (2464), (Eds.) O'Neill et al., Springer, pp. 137-144.

27. Brabazon, A., Delahunty, A., O'Callaghan, D., Keenan, P. and O'Neill, M. (2005). Financial classification using an artificial immune system, in *Business Applications and Computational Intelligence*, Voges, K. and Pope, N. (Eds.), Hershey, PA, USA: Idea Group Inc. (forthcoming).

28. Brabazon, A. and Keenan, P. (2004). A hybrid genetic model for the prediction of corporate failure, *Computational Management Science*, 1(3-4):293-310.

29. Brabazon, A. and O'Neill, M. (2003). Anticipating bankruptcy reorganisation from raw financial data using grammatical evolution, in Proceedings of EvoIASP 2003, *Lecture Notes in Computer Science (2611): Applications of Evolutionary Computing*, (Eds.) Raidl et al., Springer, pp. 368-378.

30. Brabazon, A. and O'Neill, M. (2004). Evolving technical trading rules for spot foreign-exchange markets using grammatical evolution, *Computational Management Science*, 1(3-4):311-327.

31. Brabazon, A. and O'Neill. M. (2004). Diagnosing corporate stability using grammatical evolution, *International Journal of Applied Mathematics and Computer Science*, 14(3):363-374.

32. Brabazon, A.,O'Neill, M. and Ryan, C. (2002). Grammatical evolution and corporate failure prediction, in *Proceedings of the Genetic and Evolutionary Computation Conference (GECCO 2002)*, (Eds.) Spector et al., New York, USA, July 9-13, 2002, pp. 1011-1019, Morgan Kaufmann.

33. Brabazon, A., Silva, A., Ferra de Sousa, T., O'Neill, M., Matthews, R. and Costa, E. (2005). Investigating strategic inertia using OrgSwarm, *Informatica*, 29(2):125-141.

34. Brabazon, A., Silva, A., Ferra de Sousa, T., O'Neill, M. and Costa, E. (2005). Simulating the strategic adaptation of organizations using OrgSwarm, *Handbook of Bio-inspired Algorithms and Applications*, Olariu, S. and Zomaya, A. (eds.), Chapman and Hall, pp. 303-317.

35. Brock, W., Lakonishok, J. and LeBaron B. (1992). Simple technical trading rules and the stochastic properties of stock returns, *Journal of Finance*, 47(5):1731-1764.

36. Brown, S., Goetzmann W. and Kumar A. (1998). The Dow Theory: William Peter Hamilton's track record reconsidered, *Journal of Finance*, 53(4):1311-1333.

37. Chan, L. K. C., Jegadeesh, N. and Lakonishok, J. (1996). Momentum strategies, *Journal of Finance*, 51(5):1681-1714.

38. Chao, D. and Forrest, S. (2003). Information immune systems, *Genetic Programming and Evolvable Machines*, 4(4): 311-331.

39. Chen, J. (2002). A heuristic approach to efficient production of detector sets for an artificial immune algorithm-based bankruptcy prediction system, in *Proceedings of the Congress on Evolutionary Computation 2002*, 1:932-937, New Jersey: IEEE Press.

40. Cheng, A. C. S. (1998). International correlation structure of financial market movements - the evidence from the United Kingdom and the US, *Applied Financial Economics*, 8(1):1-13.

41. Cleary, R. and O'Neill, M. (2005). An Attribute grammar decoder for the 01 MultiConstrained Knapsack Problem, in LNCS 3488 *Proceedings of the European Conference on Evolutionary Combinatorial Optimisation - EvoCOP 2005*. Lausanne, Switzerland, pp. 34-45. Springer.

42. Clerc, M. (1999). The swarm and the queen: towards a deterministic and adaptive particle swarm optimization, in *Proceedings of ICEC 1999*, pp. 1951-1957, Washington, DC.

43. Colin, A. (1994). Genetic algorithms for financial modelling, in *Trading on The Edge: Neural, Genetic and Fuzzy systems for Chaotic and Financial Markets*, Guido Deboeck (Ed.)), New York: Wiley.

44. Cover, T. (1965). Geometrical and statistical properties of systems of linear inequalities with applications in pattern recognition, *IEEE Transactions in Electron. Comput.*, EC-14:326-334.

45. Cramer, N.L. (1985). A representation for the adaptive generation of simple sequential programs, in *Proceedings of the International Conference on Genetic Algorithms and Their Applications*, pp. 183-187, Carnegie-Mellon University, Pittsburgh, PA.

46. Cybenko, G. (1989). Approximation by superpositions of a sigmoidal function, *Math. Control Signal Systems*, 2:303-314.

47. Dacorogna, M., Gencay, R., Muller, U. Olsen, R. and Pictet, O. (2001). *An Introduction to High-frequency Finance*, New York: Academic Press.

48. Dambolena, I. and Khoury, S. (1980). Ratio stability and corporate failure, *Journal of Finance*, 35(4):1017-1026.

49. Darwin, C. (1859). On the Origin of the Species by Means of Natural Selection, or the Preservation of Favoured Races in the Struggle for Life (reprinted 1985), London: Penguin Books.

50. Dasgupta, D. and Forrest, S. (1996). Novelty detection in time series data using ideas from immunology, in *Proceedings of ISCA 5th International Conference on Intelligent Systems*, Reno, Nevada, June 19-21, 1996.

51. Deboeck, G. (1994a). *Trading on the Edge: Neural, Genetic, and Fuzzy Systems for Chaotic Financial Markets*, New York: John Wiley and Sons.

52. Deboeck G. (1994b). Using GAs to optimise a trading system, in Guido Deboeck (Ed.), *Trading on The Edge: Neural, Genetic and Fuzzy systems for Chaotic and Financial Markets*, New York: John Wiley & Sons Inc.

53. DeBondt, W. and Thaler, R. (1985). Does the stock market overreact?, *Journal of Finance*, 40:793-805.

54. DeBondt, W. and Thaler, R. (1987). Further evidence on investor overreaction and stock market seasonality, *Journal of Finance*, 42(3):557-581.

55. De Castro, L. and Timmis, J. (editors). (2002). *Artificial Immune Systems: A New Computational Intelligence Approach*, London: Springer.

56. De Castro, L. and Von Zuben, F. (2002). Learning and optimization using the clonal selection principle, *IEEE Transactions on Evolutionary Computation*, 6(3):239-251.

57. Dempsey, I., O'Neill, M. and Brabazon, A. (2002). Investigations into market index trading models using evolutionary automatic programming, in *LNAI 2464, Proceedings of the 13th Irish Conference in Artificial Intelligence and Cognitive Science*, pp. 165-170, (Eds.) O'Neill et al., Springer.

58. Dempsey, I., O'Neill, M., and Brabazon, A. (2004). Grammatical constant creation, in LNCS 3103 *Proceedings of the Genetic and Evolutionary Computation Conference - GECCO 2004*, Part 2, pp. 447-458, Seattle WA, USA, Springer.

59. Dempsey, I., O'Neill, M. and Brabazon, A. (2004). Live trading with grammatical evolution, in *Proceedings of the Grammatical Evolution Workshop 2004*, a Workshop of the Genetic and Evolutionary Computation Conference, GECCO 2004.

60. Dempsey, I., O'Neill, M., and Brabazon, A. (2005). meta-grammar constant creation, in *Proceedings of the Genetic and Evolutionary Computation Conference - GECCO 2005*, pp. 1665-1672, Washington DC, USA, ACM Press.

61. Dempster, M. and Jones, C. (2001). A real-time adaptive trading system using genetic programming, *Quantative Finance*, 1:397-413.

62. Deneubourg, J., Gross, S., Franks, N. Sendova-Franks, A. Detrain, C. and Chretien, L. (1991). The dynamics of collective sorting robot-like ants and ant-like robots, *Proceedings of 1st Conference on Simulation of Adaptive Behavior: From Animals to Animats (SAB 90)*, in Meyer, J. and Wilson, S. (eds), 356-365: MIT Press.

63. Dissanaike, G. (1997). Do stock market investors overreact?, *Journal of Business Finance & Accounting (UK)*, 24(1):27-50.

64. Dorigo, M. and DiCaro, G. (1999). Ant colony optimization: a new meta-heuristic, *Proceedings of CEC*, vol 2, 1470-1477: IEEE Press.

65. Dorigo, M. and Gambardella, L. (1997). Ant colonies for the travelling salesman problem, *BioSystems*, 43:73-81.

66. Dorigo, M., Maniezzo, V. and Colorni, A. (1996). Ant system: optimization by a colony of cooperating agents, *IEEE Transactions on Systems, Man, And Cybernetics - Part B: Cybernetics*, 26(1):29-41.

67. Easterbrook, F. (1990). Is corporate bankruptcy efficient?, *Journal of Financial Economics*, 27(2):411-417.

68. Eberhart, R. Dobbins, R. and Simpson, P. (1996). *Computational Intelligence PC Tools*, Boston, MA: Academic Press.

69. Edelman, D. (2001). *The Compleat Horseplayer*, De Mare Consultants, Australia.

70. Efron, B. and Tibshirani, R. (1993). *An Introduction to the Bootstrap*. New York: Chapman and Hall.

71. Elman, J. (1990). Finding structures in time, *Cognitive Science*, 14:179-211.

72. Fahlman, S. (1988). Faster-learning variations on back-propagation: An empirical study, (Eds.) T. J. Sejnowski, G. E. Hinton and D. S. Touretzky, *1988 Connectionist Models Summer School*, San Mateo, CA, 1988. Morgan Kaufmann.

73. Fama, E. (1970). Efficient capital markets: a review of theory and empirical work, *Journal of Finance*, Vol. 25(2):383-417.

74. Fama, E. (1998a). Market efficiency, long-term returns, and behavioral finance, *Journal of Financial Economics*, 49(3):283-306.

75. Fama, E. (1998b). Efficiency survives the attack of the anomalies, *GSB Chicago Alumni Magazine*, (Winter):14-16.

76. Fernandez-Rodriguez, F., Gonzalez-Martel, C. and Sosvilla-Rivero, S. (2000). On the profitability of technical trading rules based on artificial neural networks: Evidence from the Madrid stock market, *Economics Letters*, 69:89-94.

77. Ferris, S., Jayaraman, N. and Makhija, A. (1996). The impact of Chapter 11 filings on the risk and return of security holders, 1979-1989, *Advances in Financial Economics*, 2:93-118.

78. Fitzpatrick, P. (1932). *A Comparison of the Ratios of Successful Industrial Enterprises with Those of Failed Companies*, Washington: The Accountants' Publishing Company.

79. Fogel, D. (2000). *Evolutionary Computation: Towards a New Philosophy of Machine Intelligence*, New York: IEEE Press.

80. Forrest, S., Perelson, A. Allen, L. and Cherukuri, R. (1994). Self-nonself discrimination in a computer, in *Proceedings of the 1994 IEEE Symposium on Research in Security and Privacy*, IEEE Computer Society Press: Los Alamitos, California, 1994, pp. 202-212.

81. Foster, J.A. (2001). Evolutionary computation, *Nature Genetics Reviews*, Vol. 2, pp. 428-436, June, 2001.

82. Franses, P. and Van Homelen, P. (1998). On forecasting exchange rates using neural networks, *Applied Financial Economics*, 8:589-596.

83. Friedberg, R.M. (1958). A Learning Machine: Part 1. *IBM J. Research and Development*, 2(1):2-13.

84. Friedberg, R.M., Dunham, B., North, J.H. (1959). A learning machine: Part 2. *IBM J. Research and Development*, 3:282-287.

85. Froot, K. and Thaler, R. (1990). Anomalies: foreign exchange, *Journal of Economic Perspectives*, 4(3):179-192.

86. Gately, E. (1996). *Neural Networks for Financial Forecasting*, New York: Wiley.
87. Gencay, R., Ballocchi, G., Dacorogna, M., Olsen, R. and Pictet, O. (2002). Real-time trading models and the statistical properties of foreign exchange rates, *International Economic Review*, 43(2):463-491.
88. Gencay, R., Dacorogna, M., Olsen, R. and Pictet, O. (2003). Foreign exchange trading models and market behavior, *Journal of Economic Dynamics & Control*, 27:909-935.
89. Gentry, J., Newbold, P. and Whitford, D. (1985). Classifying bankrupt firms with funds flow components, *Journal of Accounting Research*, 23(1):146-160.
90. Glassman, R. (1973). Persistence and loose coupling in living systems, *Behavioral Science*, 18:83-98.
91. Goldberg, D.E. (1989). *Genetic Algorithms in Search, Optimization and Machine Learning*, Boston: Addison-Wesley.
92. Goldberg, D.E. (2002). *The Design of Innovation: Lessons from and for Competent Genetic Algorithms*, Kluwer Academic Publishers.
93. Goldsby, R., Kindt, T., Kuby, J. and Osborne, B. (2002). *Immunology* (5th ed.), New York: W. H. Freeman & Co.
94. Gomez, F. and Miikkulainen, R. (1997). Incremental evolution of complex general behavior, *Adaptive Behavior*, 5:317-342.
95. Gomez, F. (2003). *Robust Non-linear control through neuroevolution*, PhD thesis, University of Texas at Austin, Department of Computer Sciences Technical Report AI-TR-03-303.
96. Gonzalez, F. and Dasgupta, D. (2003). Anomaly detection using real-valued negative selection, *Genetic Programming and Evolvable Machines*, 4(4):383-403.
97. Gonzalez, L. and Cannady, J. (2004). A self-adaptive negative selection approach for anomaly detection, in *Proceedings of the Congress on Evolutionary Computation 2004*, 2: 1561-1568, New Jersey: IEEE Press.
98. Goonatilake, S. and Treleaven, P. (1996). *Intelligent Systems for Finance and Business*, Chichester, UK: John Wiley and Sons.
99. Gruau, F. (1994). *Neural Network Synthesis Using Cellular Encoding and the Genetic Algorithm*, PhD Thesis, L'Ecole Normale Superieure de Lyon, l'University Claude Bernard-Lyon.
100. Gudise, V. and Venayagamoorthy, G. (2003). Comparison of particle swarm optimization and backpropagation as training algorithms for neural networks, *Proceedings of the 2003 IEEE Swarm Intelligence Symposium* (SIS '03), 110-117, IEEE Press.
101. Gurney, K. (1997). *An introduction to Neural Networks*, London: University College London Press.
102. Hagan, M. and Menhaj, M. (1994). Training feedforward networks with the Marquardt algorithm, *IEEE Transactions on Neural Networks*, 5(6):989-993.
103. Hair, J., Anderson, R., Tatham, R. and Black, W. (1998). *Multivariate Data Analysis*, Upper Saddle River, New Jersey: Prentice Hall.
104. Hambrick, D. and D'Aveni, R. (1988). Large corporate failures as downward spirals, *Administrative Science Quarterly*, 33:1-23.
105. Hancock, P. (1992). Genetic algorithms and permutation problems: a comparison of recombination operators for neural net structures, in *Proceedings of COGANN-92 workshop*, 108-122, IEEE Press.
106. Hoai, N.X., McKay, R.I.and Abbass, H.A. (2003). Tree adjoining grammars, language bias, and genetic programming, in LNCS 2610 *Genetic Programming, Proceedings of EuroGP 2003*, pp. 340-349, Springer.

107. Hofmeyer, S. and Forrest, S. (2000). Architecture for an artificial immune system, *Evolutionary Computation*, 8(4), 443-473.

108. Holland, John H. (1975). *Adaptation in Natural and Artificial Systems*, Michigan: University of Michigan Press.

109. Hong, H., Lim, T. and Stein, J. (1999). Bad news travels slowly: size, analyst coverage, and the profitability of momentum strategies, *Journal of Finance*, 55(1):265-295.

110. Hornik, K., Stinchcombe, M. and White, H. (1990). Multi-layered feedforward neural networks are universal approximators, *Neural Networks*, 2, 359-366.

111. Horrigan, J. (1965). Some empirical bases of financial ratio analysis, *The Accounting Review*, July 1965, 558-568.

112. Hu, M., Zhang, G., Jiang, C. and Patuwo, E. (1999). A cross-validation analysis of neural network out-of-sample performance in exchange rate forecasting, *Decision Sciences*, 30(1):197-216.

113. Hu, X., Shi, Y. and Eberhart, R. (2004). Recent advances in particle swarm, in *Proceedings of CEC 2004*, pp. 90-97, Portland, Oregon, 19-23 June, 2004, IEEE Press: New Jersey.

114. Iba H. and Nikolaev N. (2000). Genetic programming polynomial models of financial data series, in *Proc. of CEC 2000*, 1459-1466, IEEE Press.

115. Jamal, A. and Sundar, C. (1999). Modelling exchange rates with neural networks, *Journal of Applied Business Research*, 14(1):1-5.

116. Janeway, C., Travers, P., Walport, M. and Shlomchik, M. (2004). *Immunobiology* (6th ed.), New York: Garland Publishing.

117. Ji, Z. and Dasgupta, D. (2004). Augmented negative selection algorithm with variable-coverage detectors, in *Proceedings of CEC 2004*, pp. 1081-1088: IEEE Press.

118. Kahya, E. and Theodossiou, P. (1996). Predicting corporate financial distress: A time-Series CUSUM methodology, *Review of Quantitative Finance and Accounting*, 13:71-93.

119. Kamich, B. (2003). *How Technical Analysis Works*, New York: New York Institute of Finance.

120. Kantschik, W. and Banzhaf, W. (2002). Linear-graph GP—A new GP Structure, in LNCS 2278 *Genetic Programming, Proceedings of the 5th European Conference, EuroGP 2002*, pp. 83-92, Springer.

121. Kantschik, W., Dittrich, P., Brameier, M. and Banzhaf, W. (2002). Meta-evolution in graph GP, in LNCS 1598 *Genetic Programming, Proceedings of EuroGP 99*, pp. 15-28, Springer.

122. Kaufman, P. (1998). *Trading Systems and Methods* (3rd ed.), New York: John Wiley & Sons.

123. Kendall, M. (1953). The analysis of economic time series (part 1), prices, *Journal of the Royal Statistical Society*, 96, pp. 11-25.

124. Kennedy, J. and Eberhart, R. (1995). Particle swarm optimization, *Proceedings of the IEEE International Conference on Neural Networks*, December 1995, pp. 1942-1948.

125. Kennedy, J. and Eberhart, R. (1997). A discrete binary version of the particle swarm algorithm, *Proceedings of the Conference on Systems, Man and Cybernetics*, pp. 4104-4109: IEEE Press.

126. Kennedy, J., Eberhart, R. and Shi, Y. (2001). *Swarm Intelligence*, San Mateo, California: Morgan Kaufman.

127. Kitano, H. (1990). Designing neural networks using genetic algorithms with graph generation, *Complex Systems*, 4:461-476.

128. Kohonen, T. (1982). Self-organized formation of topologically correct feature maps, *Biological Cybernetics*, 43:59-69.

129. Kohonen, T. (1990). The Self-organizing map, *Proceedings of the IEEE*, 78(9):1464-1480.

130. Kohonen, T. (1998). The SOM Methodology, *Visual Explorations in Finance with Self-organizing Maps*, (Eds.) Deboeck, G. and Kohonen, T., pp. 159-167, Berlin: Springer-Verlag.

131. Koza, J. (1992). *Genetic Programming*, Massachusetts: MIT Press.

132. Koza, J. (1994). *Genetic Programming II*, Massachusetts: MIT Press.

133. Koza, J. R., David Andre, Bennett III, F. H. and Keane, M. (1999). *Genetic Programming III: Darwinian Invention and Problem Solving*, Morgan Kaufman.

134. Koza, J. R., Keane, M. A., Streeter, M. J., Mydlowec, W., Yu, J. and Lanza, G. (2003). *Genetic Programming IV: Routine Human-Competitive Machine Intelligence*, Kluwer Academic Publishers.

135. Kumar, N., Krovi, R. and Rajagopalan, B. (1997). Financial decision support with hybrid genetic and neural based modelling tools, *European Journal of Operational Research*, 103(2):339-349.

136. Langdon, W.B. and Gustafson, S. (2005). Genetic programming and evolvable machines: five years of reviews, *Genetic Programming and Evolvable Machines*, Vol. 6, No. 2.

137. Langdon, W.B., Gustafson, S. and Koza, J.R. (2004). The Genetic Programming Bibliography, http://liinwww.ira.uka.de/bibliography/Ai/genetic.programming.html.

138. Langdon, W.B. and Poli, R. (2002). *Foundations of Genetic Programming*, Springer-Verlag.

139. Levich, R. and Thomas, L. (1993). The significance of technical trading-rule profits in the foreign exchange market: a bootstrap approach, *Journal of International Money and Finance*, 12:451-474.

140. Levinthal, D. (1991). Random walks and organizational mortality, *Administrative Science Quarterly*, 36(3):397-420.

141. Lewin, B. (2000). *Genes VII*, Oxford University Press.

142. Lintner, G. (1998). Behavioral finance: why investors make bad decisions, *The Planner*, 13(1):7-8.

143. Lo, A. and MacKinlay, C. (1999). *A Non-random Walk Down Wall Street*, Princeton, New Jersey: Princeton University Press.

144. Lo, A. W., Mamaysky, H. and Wang, J. (2000). Foundations of technical analysis: computational algorithms, statistical inference, and empirical implementation, *Journal of Finance*, 55(4):1705-1765.

145. Lui, Y. and Mole, D. (1998). The use of fundamental and technical analyses by foreign exchange dealers: Hong Kong evidence, *Journal of International Money and Finance*, 17:535-545.

146. Lumer, E. and Faieta, B. (1994). Diversity and adaptation in populations of clustering ants, *Proceedings of Third International Conference on Simulation of Adaptive Behaviour*, pp. 501-508.

147. Manly, Bryan F J. (1994). *Multivariate Statistical Methods*, London: Chapman & Hall.

148. Matthews, B.W. (1975). Comparison of the predicited and observed secondary structure of T4 phage lysozyme, *Biochemica et Biophysica Acta.*, 405:442-451.

149. Matzinger, P. (1994). Tolerance, danger and the extended family, *Annual Review of Immunology*, 12:991-1045.
150. Matzinger, P. (2002). The Danger Model: a renewed sense of self, *Science*, 296(5566):301-305.
151. McRobert, A. and Hoffman, R. (1997). *Corporate Collapse: An Early Warning System for Lenders, Investors and Suppliers*, Roseville: NSW, McGraw-Hill (Australia).
152. Miller, J. and Thomson, P. (2000). Cartesian genetic programming, in LNCS 1802 *Genetic Programming, Proceedings of EuroGP 2000*, pp.121-132, Springer.
153. Mitchell, M. (1996). *An Introduction to Genetic Algorithms*, Cambridge, Massachusetts: MIT Press.
154. Monson, C.K. and Seppi, K.D. (2004). The Kalman Swarm (a new approach to particle motion in swarm optimization), in LNCS 3102 *Proceedings of the Genetic and Evolutionary Computation Conference-GECCO 2004*, Part 1, pp. 140-150, Seattle WA, USA, Springer.
155. Montana, D. and Davis, L. (1989). Training feedforward neural networks using genetic algorithms, in *Proceedings of the 11th International Joint Conference on Artificial Intelligence*, pp. 762-767, Morgan Kaufman.
156. Montier, J. (2002). *Behavioural Finance: Insights into Irrational Minds and Markets*, Chichester, UK: Wiley.
157. Moody's (2000). RiskCalc For Private Companies: Moody's Default Model, http://www.riskcalc.moodysrms.com.
158. Morris, R. (1997). *Early Warning Indicators of Corporate Failure: A Critical Review of Previous Research and Further Empirical Evidence*, London: Ashgate Publishing.
159. Moulton, W. and Thomas, H. (1993). Bankruptcy as a deliberate strategy: theoretical considerations and empirical evidence, *Strategic Management Journal*, 14(2):125-135.
160. Murphy, John J. (1999). *Technical Analysis of the Financial Markets*, New York: New York Institute of Finance.
161. Neely, C., Weller P. and Dittmar, R. (1997). Is technical analysis in the foreign exchange market profitable? A genetic programming approach, *Journal of Financial and Quantitative Analysis*, 32(4):405-428.
162. Nordin, P. (1997). *Evolutionary Program Induction of Binary Machine Code and Its Applications*, PhD Thesis, Universität Dortmund am Fachbereich Informatik.
163. NYSE (2005). *Market Information-Quick Reference Sheet*, http://www.nyse.com.
164. NYSE (2005). *Market Statistics*, http://www.nyse.com.
165. Ohlson, J. (1980). Financial ratios and the probabilistic prediction of bankruptcy, *Journal of Accounting Research*, 18:109-131.
166. O'Neill, M. (2001). *Automatic Programming in an Arbitrary Language: Evolving Programs with Grammatical Evolution*, PhD thesis, University of Limerick, Ireland, 2001.
167. O'Neill, M. and Brabazon, A. (2005). mGGA: The meta-Grammar genetic algorithm, in LNCS 3447 *Proceedings of the European Conference on Genetic Programming-EuroGP 2005*, pp. 311-320, Lausanne, Switzerland, Springer.
168. O'Neill, M. and Brabazon, A. (2004). Grammatical swarm, in LNCS 3120 *Proceedings of the Genetic and Evolutionary Computation Conference-GECCO 2004*, Part 1, pp. 163-174, Seattle, WA, USA. Springer-Verlag.

169. O'Neill, M., Brabazon, A. and Adley, C. (2004). The automatic generation of programs for classification problems with grammatical swarm, in *Proceedings of the Congress on Evolutionary Computation-CEC 2004*, Vol. 1, pp. 104-110, Portland, OR, USA. IEEE.

170. O'Neill, M., Brabazon, A., Nicolau, M., McGarraghy, S., Keenan, P. (2004). πGrammatical Evolution, in LNCS 3103 *Proceedings of the Genetic and Evolutionary Computation Conference-GECCO 2004*, Part 2, pp. 617-629, Seattle, WA, USA. Springer-Verlag.

171. O'Neill, M., Brabazon, A., Ryan, C. and Collins J.(2001). Evolving Market Index Trading Rules Using Grammatical Evolution, In *Lecture Notes in Computer Science: Applications of Evolutionary Computing*, pp. 343-353, (Eds.) E. Boers et al., Berlin: Springer.

172. O'Neill, M., Cleary, R. and Nikolov, N. (2004). Solving Knapsack problems with attribute grammars, in Poli, R. et al. (Eds.) *Grammatical Evolution Workshop 2004, Proceedings of the Workshops, Genetic and Evolutionary Computation Conference GECCO 2004*. Seattle, WA, USA, June 2004.

173. O'Neill, M. and Ryan, C. (2001). Grammatical evolution, *IEEE Trans. Evolutionary Computation*, 5(4):349-358.

174. O'Neill, M. and Ryan, C. (2003). *Grammatical Evolution: Evolutionary Automatic Programming in an Arbitrary Language*, Boston: Kluwer Academic Publishers.

175. O'Neill, M. and Ryan, C. (2004). Grammatical evolution by grammatical evolution. The evolution of grammar and genetic code, *LNCS 3003. Proc. of the European Conference on Genetic Programming 2004*, pp. 138-149, Coimbra, Portugal. Springer.

176. O'Neill, M. and Ryan, C. (Eds.) (2002). Grammatical Evolution Workshop 2002, in Barry, A. (Ed.), *Proceedings of the Workshops, Genetic and Evolutionary Computation Conference GECCO 2002*, New York, NY, USA, July 2002.

177. O'Neill, M. and Ryan, C. (Eds.) (2003). Grammatical Evolution Workshop 2003, in Barry, A. (Ed.), *Proceedings of the Workshops, Genetic and Evolutionary Computation Conference GECCO 2003*, Chicago, IL, USA, July 2003.

178. O'Neill, M. and Ryan, C. (Eds.) (2004). Grammatical Evolution Workshop 2004, in Poli, R. et al. (Eds.), *Proceedings of the Workshops, Genetic and Evolutionary Computation Conference GECCO 2004*, Seattle, WA, USA, June 2004.

179. Osler, C. (2003). Currency orders and exchange rate dynamics: An explanation for the predictive success of technical analysis, *Journal of Finance*, 58(5):1791-1820.

180. Parpinelli, R. and Lopes, H. (2002). Data mining with an ant colony optimization algorithm, *IEEE Transactions on Evolutionary Computing*, 6(4):321-332.

181. Price, K. (1999). An introduction to differential evolution, in *New Ideas in Optimization*, eds. Corne, D., Dorigo, M. and Glover, F., pp. 79-108, McGraw-Hill, London.

182. Pring, M. (1991). *Technical Analysis Explained: The Successful Investor's Guide to Spotting Investment Trends and Turning Points*, New York: McGraw-Hill.

183. Ramos, V. and Merelo, J. (2002). Self-organized stigmergic document maps: Environment as a mechanism for context learning, *Proceedings of AEB 02*, 8-10 February 2002, Merida, Spain.

184. Refenes, A.N., Bentz, Y., Bunn, D. W., Burgess, A.N. and Zapranis A.D. (1997). Financial time series modelling with discounted least squares back-propagation, *Neurocomputing*, 14:123-138.
185. Rothlauf, F. (2002). *Representations for Genetic and Evolutionary Algorithms*, Physica-Verlag.
186. Ruggiero, M. A. (1997). *Cybernetic Trading Strategies*, New York: Wiley.
187. Rumelhart, D., Hinton, G. and Williams, R. (1986). Learning internal representations by back-propagating errors, *Nature*, 323:533-536.
188. Russel, P., Branch, B. and Torbey, V. (1999). Market valuation of bankrupt firms: is there an anomaly?, *Quarterly Journal of Business and Economics*, 38:55-76.
189. Ryan C., Collins J.J. and O'Neill M. (1998). Grammatical evolution: evolving programs for an arbitrary language, *Lecture Notes in Computer Science 1391, Proceedings of the First European Workshop on Genetic Programming*, pp. 83-95, Springer.
190. Sastry, K., Goldberg, D.E. (2003). Probabilistic model building and competent genetic programming, Riolo, R.L. and Worzel, B. (Eds.) *Genetic Programming Theory and Practice*, pp. 205-220, Kluwer.
191. Schalkoff, R. (1992). *Pattern Recognition - Statistical, Structural and Neural Approaches*, New York: Wiley.
192. Schumpeter, J. (1934). *The Theory of Economic Development*, Cambridge, MA: Harvard Business Press.
193. Serrano-Cina, C. (1996). Self organizing neural networks for financial diagnosis, *Decision Support Systems*, 17(3):227-238.
194. Settles, M., Nathan, P. and Soule, T. (2005). Breeding swarms: A new approach to recurrent neural network training, in *Proceedings of the Genetic and Evolutionary Computation Conference (GECCO 2005)*, Beyer et al. (Eds.), Washington, USA, June 25-29, 2005, 1:185-192, ACM Press.
195. Shah, J. and Murtaza, M. (2000). A neural network based clustering procedure for bankruptcy prediction, *American Business Review*, 18(2):80-86.
196. Shan, Y., McKay, I., Baxter, R., Abbass, H., Essam, D., Nguyen, H. (2004). Grammar model-based program evolution, in *Proceedings of the 2004 IEEE Congress on Evolutionary Computation*, pp. 478-485, IEEE Press.
197. Shefrin, H. (2000). *Beyond Greed and Fear: Understanding Behavioral Finance and the Psychology of Investing*, Harvard Business School Press, Boston, USA.
198. Silva, A., Neves, A. and Costa, E. (2002). An empirical comparision of particle swarm and predator prey optimisation, in *Lecture Notes in Artificial Intelligence (2464), Proceedings of AICS 2002*, O'Neill et al. (Eds.), pp. 103-110, Springer.
199. Smith, R. and Winakor, A. (1935). Changes in the Financial Structure of Unsuccessful Corporations, *University of Illinois, Bureau of Business Research, Bulletin No. 51*.
200. Smith, T. (1992). *Accounting for Growth*, London: Century Business.
201. Spencer, H. (1864). *The Principles of Biology*, Volume 1, London and Edinburgh: Williams and Norgate.
202. Standard & Poor's (2002). Standard & Poor's Rating Services, *Statement at the US SEC Public Hearing on the Role and Function of Credit Rating Agencies in the US Securities Markets*, 15 November 2002.
203. Stanley, K. and Miikkulainen, R. (2002). Evolving neural networks through augmenting topologies, *Evolutionary Computation*, 10(2):99-127.

204. Storn, R. and Price, K. (1995). Differential evolution-a simple and efficient adaptive scheme for global optimization over continuous spaces, *Technical Report TR-95-012: International Computer Science Institute, Berkeley*, 1995.
205. Storn, R. and Price, K. (1997). Differential evolution-a simple and efficient heuristic for global optimization over continuous spaces, *Journal of Global Optimization*, 11:341-359.
206. Storn, R. (1999). System design by constraint adaptation and differential evolution, *IEEE Transactions on Evolutionary Computation*, 3:22-34.
207. Subing, Z. and Zemin, L. (2001). Neural network training using ant algorithm in ATM traffic control, *Proccedings of the IEEE International Symposium on Circuits and Systems* (ISCAS 2001), 2:157-160.
208. Sung, T., Chang, N. and Lee, G. (1999). Dynamics of modelling in data nining: interpretative approach to bankruptcy prediction, *Journal of Management Information Systems*, 16(1):63-85.
209. Svangard, N., Nordin, P., Llyod, S. and Wihlborg, C. (2002). Evolving short-term trading strategies using Genetic Programming, in *Proceedings of CEC 2002*, 2006-2010, IEEE Press.
210. Sweeney, R. (1986). Beating the foreign exchange market, *Journal of Finance*, 41(1):163-182.
211. Taylor, M. and Allen, H. (1992). The use of technical analysis in the foreign exchange market, *Journal of International Money and Finance*, 11:304-314.
212. Teller, A. and Veloso, M. (1995). PADO: A new learning architecture for object recognition, *Symbolic Visual Learning*, pp. 81-116, Oxford University Press.
213. Thaler, R. (1993). *Advances in Behavioural Finance*, New York: Russell Sage Foundation.
214. Thierens, D. (1999). Scalability problems of simple genetic algorithms, *Evolutionary Computation*, 7(4):331-352.
215. Trigueiros, D. and Taffler, R. J. (1996). Neural networks and empirical research in accounting, *Accounting and Business Research*, 26(4):347-355.
216. Vaga, T. (1990). The coherent market hypothesis, *Financial Analysts Journal*, 46(6):36-49.
217. Varetto, F. (1998). Genetic algorithms in the analysis of insolvency risk, *Journal of Banking and Finance*, 22(10):1421-1439.
218. Whigham P.A. (1996). Grammatical bias for evolutionary learning, *PhD Thesis*, University of New South Wales, Australian Defence Force Academy.
219. Wilson, N., Chong, K. and Peel, M. (1995). Neural network simulation and the prediction of corporate outcomes: some empirical findings, *International Journal of the Economics of Business*, 2(1):31-50.
220. Wilson, G.C., McIntyre, A. and Heywood, M.I. (2004). Resource review: three open source systems for evolving programs–Lilgp, ECJ and Grammatical Evolution, *Genetic Programming and Evolvable Machines*, 5(1):103-105.
221. Wong, B., Lai, V. and Lam, J. (2000). A bibliography of neural network business applications research: 1994-1998, *Computers and Operations Research*, 27:1045-1076.
222. Wong, M.L. and Leung, K.S. (2000). *Data Mining Using Grammar Based Genetic Programming and Applications*, Kluwer Academic Publishers.
223. Yao, J. and Tan, C. (2000). A case study on using neural networks to perform technical forecasting of forex, *Neurocomputing*, 34:79-98.
224. Yao, X. (1999). Evolving artifical neural networks, *Proceedings of the IEEE*, 87(9):1423-1447.

225. Zhang, J., Martin, E.B., Morris, A.J. and Kiparissides C. (1997). Inferential estimation of polymer quality using stacked neural networks, *Computers and Chemical Engineering*, 21(Supplement):1025-1030.
226. Zirilli, J. (1997). *Financial Prediction Using Neural Networks*, London: Thomson Computer Press.
227. Zmijewski, M. (1984). Methodological issues related to the estimation of financial distress prediction models, *Journal of Accounting Research-Supplement*, 59-82.

Index

Natural Computing Series

W.M. Spears: **Evolutionary Algorithms. The Role of Mutation and Recombination.**
XIV, 222 pages, 55 figs., 23 tables. 2000

H.-G. Beyer: **The Theory of Evolution Strategies.** XIX, 380 pages, 52 figs., 9 tables. 2001

L. Kallel, B. Naudts, A. Rogers (Eds.): **Theoretical Aspects of Evolutionary Computing.**
X, 497 pages. 2001

G. Păun: **Membrane Computing. An Introduction.** XI, 429 pages, 37 figs., 5 tables. 2002

A.A. Freitas: **Data Mining and Knowledge Discovery with Evolutionary Algorithms.**
XIV, 264 pages, 74 figs., 10 tables. 2002

H.-P. Schwefel, I. Wegener, K. Weinert (Eds.): **Advances in Computational Intelligence.
Theory and Practice.** VIII, 325 pages. 2003

A. Ghosh, S. Tsutsui (Eds.): **Advances in Evolutionary Computing. Theory and
Applications.** XVI, 1006 pages. 2003

L.F. Landweber, E. Winfree (Eds.): **Evolution as Computation.** DIMACS Workshop,
Princeton, January 1999. XV, 332 pages. 2002

M. Hirvensalo: **Quantum Computing.** 2nd ed., XI, 214 pages. 2004 (first edition
published in the series)

A.E. Eiben, J.E. Smith: **Introduction to Evolutionary Computing.** XV, 299 pages. 2003

A. Ehrenfeucht, T. Harju, I. Petre, D.M. Prescott, G. Rozenberg: **Computation in Living
Cells. Gene Assembly in Ciliates.** XIV, 202 pages. 2004

L. Sekanina: **Evolvable Components. From Theory to Hardware Implementations.**
XVI, 194 pages. 2004

G. Ciobanu, G. Rozenberg (Eds.): **Modelling in Molecular Biology.** X, 310 pages. 2004

R.W. Morrison: **Designing Evolutionary Algorithms for Dynamic Environments.**
XII, 148 pages, 78 figs. 2004

R. Paton[†], H. Bolouri, M. Holcombe, J.H. Parish, R. Tateson (Eds.): **Computation in Cells
and Tissues. Perspectives and Tools of Thought.** XIV, 358 pages, 134 figs. 2004

M. Amos: **Theoretical and Experimental DNA Computation.** XIV, 170 pages, 78 figs. 2005

M. Tomassini: **Spatially Structured Evolutionary Algorithms.** XIV, 192 pages, 91 figs.,
21 tables. 2005

G. Ciobanu, G. Păun, M.J. Pérez-Jiménez (Eds.): **Applications of Membrane Computing.**
X, 441 pages, 99 figs., 24 tables. 2006

K. V. Price, R. M. Storn, J. A. Lampinen: **Differential Evolution.** XX, 538 pages,
292 figs., 48 tables and CD-ROM. 2006

A. Brabazon, M. O'Neill: **Biologically Inspired Algorithms for Financial Modelling.**
XVI, 275 pages, 92 figs., 39 tables. 2006